THE ARMCHAIR ENVIRONMENTALIST

3 minute-a-day action plan to save the world

KAREN CHRISTENSEN

MQP

MQ Publications Limited
12 The Ivories, 6–8 Northampton Street, London N1 2HY
tel: 020 7359 2244, fax: 020 7359 1616
e-mail: mail@mqpublications.com
www.mqpublications.com

ISBN: 1-84072-624-5

10 9 8 7 6 5 4 3 2 1

Printed in Spain

Text and cover printed on
100% recycled paper.

CONTENTS

FOREWORD BY JONATHON PORRITT

While our awareness of green issues has greatly increased, driven largely by a constant flow of powerful scientific evidence, our efforts to redress negative impacts on the environment have been patchy and wholly inadequate. Politicians still don't see how they can gain electoral advantage out of an environment agenda, and the response of most businesses has been slow and unimaginative.

This leaves a lot of people worried about whether or not they themselves can make a difference as individuals. And here the picture is more encouraging. To be sure, there's still plenty of confusion around, but more and more people now readily accept that we all have a slice of today's environmental action. As Karen Christensen keeps reminding us, there's a limit to what the politicians and businesses can do if we – as citizens and consumers – aren't prepared to do our bit. "If we all do a little, it adds up to a lot."

And in future years we have to hope for a rather more powerful impact not just on people's behaviour, but on society's values. In that respect, it's the celebratory element in *The Armchair Environmentalist* that I find so inspiring. There really is no point being pious about seeking out a more environmentally and socially responsible lifestyle. At its simplest (but often overlooked) level, environmentalism is all about celebrating the gift of life – including a living relationship with the rest of life on Earth. Better by far to celebrate that gift in joy rather than garbed in sackcloth and ashes.

There's an interesting conundrum at work here. Even though it's true that these days we all agree that learning to live more sustainably on this planet is absolutely necessary, that doesn't necessarily make us feel good about it. Indeed, the more successful in demonstrating that necessity that scientists and campaigners are, the less desirable it somehow seems to become. Necessity may well be the mother of invention, but if desire is the real driver of human behaviour and creativity, then the necessary has to be made desirable before any kind of transformation becomes possible.

I fear that the genius and the creativity required to make it desirable would appear still to be in somewhat short supply. Just compare the often ludicrous levels of hyped-up excitement generated around so-called 'technological breakthroughs' – in biotech, nanotech, infotech, or indeed any other 'tech' that you like to think of – with the rather downbeat accounts of the need for things such as energy efficiency, renewable technologies, pollution control and waste management. Psychologically, it sometimes seems as if we're on a hiding to nothing!

Ultimately, it all comes down to what it is that makes us feel good about life. And the odd thing about the last 30 years or more of breakneck economic growth and consumption-driven affluence is that it hasn't led to corresponding increases in personal well-being and happiness. Which is precisely why Karen Christensen's own secret to happiness ("not getting more, but wanting less") provides such a fitting foundation for her words of wisdom.

INTRODUCTION

"Save the planet? Who, me?"

I'm not a professional environmentalist. I'm a busy entrepreneur and occasional author, and I started writing about saving the planet because when I was a new mother I looked for information and couldn't find anything that worked for me. I needed practical ideas that wouldn't take a lot of time or cost a lot of money.

You and I can't start a wind farm or set up a government committee or legislate taxes, though we should be encouraging these activities. We read about environmental problems or we see them right in front of us, in our neighbourhoods and as we go to work, and we want to be able to do something right now, in just a few minutes, that will make the world a better place. We really want to make a difference – even if it's a small one – and we want to show that we care.

That's what this book is about, the hundreds of little things we can do to help protect the beautiful planet we live on. These changes are not trivial. If we all do a little, it adds up to a lot.

By giving a few minutes a day to changing the way you interact with your personal bit of the global environment, you're helping to reverse direction, giving the planet a chance to regroup. You're also helping design sustainable lifeways and working with millions of people around the world who care about the natural world and a healthy future for all of us, and for our children.

Everybody knows that there are changes we can make at home to help protect the earth, but many of them seem too complicated or daunting – or just plain trivial. Like you, I want maximum impact for the time I put in and wanted to be able to do a little here and there. It isn't that I don't like convenience. But I also like food that really tastes like something. I like to use skin cream made with ingredients that don't cause birth defects or kill off fragile ecosystems. And I like being able to enjoy a sunny day without worrying about skin cancer.

This isn't easy, I know. We are constantly bombarded with advertising that tells us about all the things we need and ought to want. Here's a suggestion: think about the times in your life when you've been most content, the times of true joy and fulfillment. How much, really, does all that 'stuff' actually matter?

Most environmental books start with recycling, but in planetary terms it's our use of electricity and other energy and our use of fresh water that has the greatest impact. There are six billion people on the planet and most of them would like to live like folks in the urban west. We're polluting the environment with toxic chemicals and altered genetic material, and we're spreading: housing and commercial developments sprawl across vast areas that used to be habitat for native plants and animals. Our invasive behaviour makes it hard – even impossible – for other species to survive.

You and I, people living in the developed world and urban areas, do far more than our share of damage. The upside, though, is that we have lots of opportunity to change, and that's what this book is all about!

KAREN CHRISTENSEN (karen@berkshirepublishing.com)

CHAPTER ONE

NO PLACE LIKE HOME:
green home, green planet

We've been living on Planet Earth as if we were making a one-night stop at a cheap hotel. We throw towels all over the place, pile up the take-away food cartons and pop bottles, and don't worry about the hairs in the bathtub or hairspray on the mirror.

But our planet isn't a temporary stop, and there's no maid to tidy up in the morning. It's true that Nature can clean up almost anything, given enough time, but natural cycles often take centuries, or longer.

We'll grow old with the problems our parents started – problems we've been making worse. Wouldn't it be nice to think that when we have children or grandchildren they won't grow up wearing masks to school and that there will still be butterflies fluttering around the flowers in the garden? To make that happen, we need to realize that the Earth is our home, not a temporary stop.

Learn to live sustainably

We know that our activities have made massive, and threatening, changes to the planet. The core of environmental thinking is the question of how we humans can live – successfully and sustainably – on this beautiful planet of ours. Sustainability means making ourselves at home on Earth in a way that works today and will go on working in the years ahead. The best place to start saving the planet is in our own homes.

Home: the heart of change

Building construction is booming all over the developed world, and houses are becoming ever bigger guzzlers of materials and energy. Activities related to building are responsible for 35 to 45 per cent of carbon dioxide releases, contributing to global warming and stratospheric ozone depletion.

As well as that, buildings absorb 75 per cent of all the PVC – a plastic associated with health and environmental problems – that's manufactured, 25 per cent of the virgin wood that's cut and milled, and 40 per cent of the stone, gravel, sand, and steel that we use. Additionally, construction and demolition generate 25 per cent of the municipal solid waste stream.

And that's just to get them built. After that, the impact continues: buildings use 40 per cent of total energy, most of which still comes from fossil fuels; 67 per cent of electricity; and 16 per cent of water. Turning down the heat and insulating the attic may seem mundane or quirky, but when you think about these figures it's obvious that these steps truly matter. One piece of good news is that architects

are increasingly conscious of green issues and are using new approaches that can easily cut domestic energy use by three quarters. Such practices as using advanced insulation materials, installing air-to-air heat exchangers that expel stale air from super-insulated buildings while preheating incoming fresh air, and carefully orienting windows all make a huge difference.

For our part, we can buy energy-efficient appliances and products and choose smaller houses. Big houses use more energy and materials to build and maintain, and housing developments are, along with roads, the major cause of species loss. Only 55 per cent of new homes are built on previously developed land. If you're moving, look for a place where you won't be dependent on a car and where you'll have space to grow some of your own food, either in your own garden or in a nearby community plot. Consider these things, too:

- closeness to schools, shops, libraries, parks and facilities.
- availability of public transport, including buses, trains, and cycle paths.
- local councils or neighbourhood groups advocating progressive environmental policies and services.

Did you know? Better draughtproofing can increase indoor air pollution, so it's vital to reduce sources of indoor toxins at the same time as you take steps to reduce energy losses.

YOUR CARBON CONTRIBUTION

Roughly a third of CO_2 emissions come from industry, and a third from transportation. The remaining third is divided between the commercial sector and residences. Every time we switch on a light or surf the Internet, we're creating pollution and adding to the gases that contribute to global warming. And every product we buy contains 'embodied energy' – the energy required first to make it and then to get it to you. On average, each of us is responsible for some 50 tons of carbon dioxide every year!

Here's how to cut down that number:

- Choose the smallest home possible and share it or rent out excess space.
- Set your thermostat at no more than 16°C. It's much healthier not to live in a hothouse. Wear beautiful sweaters and shawls.
- Don't heat unused rooms.
- Buy certified organic produce as it is grown without energy-intensive chemicals.
- Drink tap water. Transporting mineral waters around the world creates pollution and adds to global warming.
- Plant lots of trees, shrubs, flowers and vegetables – plants absorb carbon dioxide.
- Choose new appliances that have an energy rating label: the more stars the more energy efficient
- Install compact fluorescent light bulbs.
- Choose a greener electric company that uses renewable energy sources – more options are coming all the time.

- Install ceiling fans. Using very little electricity, they cool a house in summer and circulate warm air in the winter.

Use the sun

- Redefine the way you use your space by making sure to live in the sunniest rooms during the winter.
- If you're building, put in large windows on the south side of the building and smaller ones on the north.
- Leave curtains open during the day and close them at night
- Dark colours absorb sunlight and heat, light colours reflect it – so decorate accordingly.
- Build a conservatory – it's a true sun trap.

Keeping cool

- Wear loose, light clothing and avoid wearing synthetic fabrics.
- Use a pretty hand fan.
- Regular sprays with spring water will keep you cool.
- Open windows at night and close them during the day.
- Keep the curtains closed and shades drawn to prevent direct sunlight from shining into the house. Shade windows by growing climbing plants outside.
- Open windows to allow cross-ventilation, or use an electric fan.
- Instead of air-conditioning, install a large fan in the attic. Run it in the evening to pull cool air through the house.
- If you have air-conditioning, ask about its energy-efficiency rating and ways to micro-seal the ducts. Use it only in extreme conditions, not to refrigerate the house.

Lighting

The familiar modern light bulbs, called 'incandescent' because of the filament, waste energy. Compact fluorescent light bulbs produce the same amount of light for about a quarter of the electricity, and last eight times longer. Using them saves an astonishing amount of money over a couple of years – and is probably the easiest thing you can do for the planet. They do cost more initially, but their low energy use means that you will save money, as well as energy, in the long run.

Imagine... If every household in the UK changed just two of their regular bulbs for energy efficient ones, the energy saved would power all the street lighting in the UK.

INDOOR AIR POLLUTION

Our bodies are designed for life in the great outdoors, but if you are like the average person in the developed world, you spend fully 90 per cent of your time inside, where the air is two to five times more polluted than outside!

This invisible pollution comes from volatile organic compounds, or VOCs. These are mainly produced by chipboard, MDF, soft plastics, plastic foams, caulking and sealing materials, paints and varnishes, office machines, and cleaning and personal care products. Commercial air fresheners, moth balls, solvents and aerosol sprays of any type are particularly hazardous.

Contrary to expectation, studies have found that colds and 'flu are less likely in draughty buildings. This may be because viruses build up in well-sealed buildings or because our immune systems need fresh air to function effectively. Exposure to VOCs can result in dizziness, headaches and nausea, can cause or exacerbate respiratory complaints such as asthma, and suppresses the immune system by reducing the production of lymphocytes and antibodies. To top it off, using a gas stove or an aromatherapy candle can produce carbon monoxide and particulate levels as high as those experienced in heavy traffic! But it's easy to make changes to make your indoor environment clean and green.

Breathing sweetly

Commercial air fresheners cover unpleasant odours with chemical scents. Some go further, coating your delicate nasal passages with oily film, or numbing them with nerve-deadening compounds. There's no need for them in your green home!

- Open the windows for a daily airing – this will circulate air and clear offensive smells and any toxic VOCs that might have built up. But don't open the windows while you're heating!

- Look into installing an air-to-air heat exchanger, because they save energy costs while they bring in fresh, outdoor air.
- Empty the rubbish bin frequently. If you're composting kitchen waste separately, the bin will stay dry and clean.
- Avoid all aerosols by choosing products which have a pump action instead.
- All indoor plants improve air quality. The common spider plants (*Chlorophytum*) are particularly good at reducing indoor toxins.
- Fill bowls with a natural pot pourri or with whole cloves and cinnamon bark. Alternatively, fill a spray bottle with water mixed with a lavender oil to use as a spray freshener.

Cleaning tips

Household cleaners do a lot of damage to the environment simply because we use so many of them. They are also a major contributor to indoor air pollution.

- Check out the latest green cleaning products.
- Put doormats outside and inside to trap incoming dirt.
- Make it a household rule that shoes come off at the door, as they do in Japan.
- Use water to soak off grime – it's the first and best cleaner.
- Read warning labels and store and use products carefully. Even natural cleaners can be skin irritants and should be kept out of children's reach.

Germ-free living

A healthy human immune system can easily cope with normal household germs and is, in fact, at greater risk from the chemicals in many cleaning products. Germicidal items such as kitchen sponges and teething toys are not necessary in the green home. They, like commercial disinfectants, contain chemicals which can affect the central nervous system and cause organ damage, as well as mimicking human hormones.

These products will also alter the balance of microbial life in the home, leaving behind those that are harder to kill. Surviving bacteria are likely to become resistant to antibiotics – so our homes, like hospitals, may harbour truly dangerous bacteria. Try natural disinfectants, which include borax, tea tree oil, grapefruit seed oil, and other citrus extracts. If you do need a stronger disinfectant, use a properly diluted solution of chlorine bleach.

You are unlikely to have a problem with insect pests if you pay extra attention to cleaning and tidying and eliminate potential entrances for them. Store food in airtight containers such as glass jars. Flies dislike mint, eucalyptus, citrus, citronella, and clove oils. You can trap ants in a small jar of sugar solution – cover the top with a bit of kitchen foil and poke a small hole in it. They'll scale the glass to find the sugar. Not pretty, but it's very effective.

Did you know? Dishwashers with powerful commercial washing powder create a plume of toxic vapour.

SWAP SHOP

Living well on Earth means keeping possessions to a reasonable, sustainable level, and thinking about what you really need; what will truly add to your quality of life. Environmentalists have always promoted second-hand shopping and bartering as ways to reduce our overall consumption. Online services and bulletin boards can be a great way to find what you need in 'gently used' condition.

Better yet, help set up a local neighbourhood bulletin board or swap meet to help local trading. We need ways to advertise freebies too – your unwanted cupboard may be just what I need, and there are many people in every community who could make good use of things others toss out. In fact, it may well be more important to buy recycled products than to recycle items yourself, because successful recycling depends on consumer demand. Choose things made from 'post-consumer' paper or plastic.

Living room

Who would have thought that an ordinary armchair could be so dangerous? But it turns out that PBDE, a fire retardant used in foam cushions and mattresses, is turning up in our bodies – and in women's breast milk. If you're shopping for furniture, look for PBDE-free sofas, chairs and beds. There's a surprising amount of chemical finishing used on the things we bring into our homes and that's what causes the strong 'new house' smell.

These chemicals eventually evaporate, but they add a chemical burden to our bodies that can make allergies and asthma worse and cause what's known as 'sick building syndrome'.

Fortunately, there is a wide array of green alternatives, from organic cotton upholstery to unfinished, hand-washable throw rugs and carpets. Get out there – or go online – and explore!

Fitted carpets are a special problem for anyone with allergies. They trap dust mites, flea eggs and moulds and release VOCs from glues and finishing. Plain wooden floors and throw rugs that can be washed – or hung on a line and beaten – are better choices.

The first place to look for green home furnishings is in secondhand and antique markets. Reusing older pieces reduces the demand for new wood and other resources. The good news is that you can buy beautiful old rugs for less than new ones and they will only increase in value.

Ask about the wood used in any new furniture you buy. 'Certified wood' is produced to standards set by the Forest Stewardship Council. 'Rediscovered wood' is salvaged from demolished properties, urban tree salvage and even demolition landfills. 'Green wood' includes formaldehyde-free composite panels, arsenic-free pressure-treated lumber, engineered structural wood products and even 'plastic lumber' made from recycled plastics. Choose local woods in preference to tropical hardwoods, such as iroko and mahogany, because even if these are sustainably harvested, transporting them takes a lot of energy.

Green decorating

Eco living has a comfortable simplicity that's easy to maintain – uncluttered spaces, plenty of mirrors for extra brightness, green plants and an orientation to the outdoors.

Here's how to achieve a relaxed living style:

- Decorate with bright, energising colours.
- Choose simply-styled, affordable, sustainably-produced woods and other natural materials.
- Introduce lots of glass, mirrors and furnishings designed to allow light to flow.
- Buy hand-made regional tiles and furniture to promote local economies and reduce transportation costs.
- Hunt down furnishings that serve multiple uses.
- Use simple household tools such as wooden clothes racks, roller towels, carpet sweepers, straw brooms, aprons.

Did you know? Cotton is one of the world's most polluting crops. Farmers worldwide spend £1.4 billion a year on pesticides to protect the crop from insects and it uses excessive amounts of the world's fresh water. The solution is to choose organic cotton, or other fabrics if appropriate. Remember that some rayon is made from eucalyptus trees that are both ecologically and economically damaging to the countries, such as India, where they are grown.

MAKE YOUR KITCHEN GREENER

The kitchen is the heart of our homes – so here are tips for reducing waste, using less energy and water – and greening your garden at the same time, too!

- Use a ceramic or stainless steel waste container with a lid that can easily be washed or sterilized with boiling water.
- Reuse a large plastic tub with a cover that you can recycle when it gets old. It should be large enough to hold a couple of days' vegetable scraps – you don't want to be trekking outside several times a day to empty it.
- Pare and prepare vegetables on a sheet of newspaper. Then gather up the edges and put the whole lot in the compost bin.
- Use non-electrical equipment whenever possible. Look for inexpensive and attractive coffee grinders, rotary beaters and cutting boards. Blenders and food processors are great, but chopping by hand is a pleasurable skill.
- Cook in the microwave, which uses only a fraction of the energy required by other cooking appliances. Microwaves are great for reheating homemade dishes as well as for steaming vegetables and fish.
- Run full dishwasher loads on an economy cycle – this takes less water than washing by hand.
- Scrape rather than rinse dishes for the dishwasher and don't waste water washing cans and bottles for recycling. You need to do just enough to get rid of smells and discourage ants, flies and other insects.

Watch those disposables

You'll cut kitchen waste by 25 per cent if you compost. Next, try to cut your packaging and disposables by 25 per cent, too. Here are some suggestions:

Paper towels: Use terrycloth towels for your hands and a dishcloth for the counter tops.

Sponges: Cotton or linen dishcloths last longer than sponges.

Disposable cleaning cloths: Cut them from old sheets and shirts. Linen is best for windows.

Paper napkins: Use cloth napkins made of sturdy cotton fabric that does not need to be ironed.

Kitchen foil and clingfilm: Cover bowls with a side plate or saucer, and casseroles and gratin dishes with their own covers or small cookie sheets. Reuse foil and take-away containers.

Plastic bags: Wash and reuse. Dry over an empty bottle, or make a 'bag bouquet' by standing long straight twigs in a slim vase – hang the washed plastic bags over the ends of the twigs.

Coffee filters: Replace paper filters with a reusable gold or hemp filter, or a cafetière. Use loose tea instead of teabags.

Paper and polystyrene cups and paper plates: Keep pretty flea market plates and cutlery in a basket, ready for impromptu picnics.

Plastic rubbish bags: Your rubbish will be dry if you compost your food scraps, so you won't need plastic bags any more!

Carrier bags: Plastic bags actually have less impact on the environment than heavy paper bags, but in either case it's far better to bring your own canvas or string bags.

MAKE YOUR BATHROOM GREENER

Fresh water shortages are going to be one of the top environmental challenges this century because ancient aquifers – the reservoirs of underground water that feed wells and springs – are being drained to supply our incessant agricultural, industrial and consumer demand. Reducing water use is just as important as recycling, and there are plenty of ways to make your bathroom green.

- Toilets use more water than any other home device. Put water-savers in the cistern or install one of the modern low-flow models.
- Shower instead of bathe. Showers use less water than baths – up to ten times less, depending on how long you take.
- Install a low-flow shower head and aerator attachment.
- Use biodegradable soaps and bath products.
- Consider ways to get water from the bath or shower to the garden – look online for ideas about 'grey-water systems'.

MAKE YOUR BEDROOM GREENER

Although more people say they are short of sleep, we aren't sleeping less. We're just not getting the rest we need, and that's a reflection of the psychological impact of our hectic way of life. Furnish your bedroom for tranquillity, with hand-me-downs, quaint charity or junk shop finds and new pieces made from sustainably-harvested woods such as beech, birch, oak and pine. There's no need to throw out what you have now, but if you buy a new bed you can choose a green mattress – made from organic wool – some hypo-allergenic pillows and bedding made from untreated organic cotton. The most beautiful organic sheets are undyed and made from different colours of natural cotton. Linen is another gorgeous fabric – expensive, but it will outlast cotton by over 10 years. Or consider hemp sheets, which look and feel like linen and wear well.

Reduce your 'electric load'

In the future, we'll see houses designed to minimize our exposure to electro-magnetic fields, but in the meantime try these ideas:

- Minimize the electrical equipment in your bedroom. Use a hot water bottle or grain-bag heated in the microwave instead of an electric blanket.
- Keep all electrical equipment and flex at least a metre from your bed. A ceiling or wall-mounted light is best.
- Turn off and unplug equipment such as TV sets, VCRs and DVDs. Do you really need these in your bedroom anyway?
- If you have to keep your mobile phone with you, put it as far as possible from your reproductive organs and head.
- Choose non-ionizing smoke detectors.

A GREEN WASH

Green laundry products have caused more disappointment than anything else people have tried for the sake of the environment, but don't give up – they're getting better all the time and hi-tech mainstream products don't do all they promise anyway. The key to greener washing is to make a mental division between the clothes that can be washed in any manner – everyday black pants, for example – and and those that need special attention, such as white shirts. Don't buy detergents that are full of additives, enzymes and fragrances. These will pollute water supplies, damage animals and cause allergic reactions.

Research has shown that plain water and ordinary washing soda gets most clothes just as clean as detergents in hard water areas. Detergents simply loosen grime, and water does the same, especially if you let clothes soak for ten to 15 minutes. Keeping whites really bright does require special treatment – whether it's bleach of some kind or sunshine – so make your life easier by choosing natural-coloured linen and darker colours when possible. You'll find details on the best ways of handling stains and fabrics that require dry-cleaning on page 28.

Energy saving

- Use a front-loading washing machine; they use less water and energy and are cheaper to run.
- Wash full loads, with cycles as short and cool as possible.
- Hang clothes outside to dry or on racks indoors.
- When you use a tumble-drier, run loads back to back, while the drum is still hot.

Stain removal

Many clothes are discarded because of stains, so learn some basics about removing them. Have one of the new and excellent non-toxic products on hand – fast treatment is the most effective way to remove most stains.

- Try cold water first on protein-based spots such as blood, egg or gravy. Hot water will set the stain.
- Pour salt immediately on red wine or fruit stains, and brush it off as it draws out the stain.
- Try one part borax to eight parts water for blood, chocolate, coffee, or tea.
- Boiling water is the traditional method for fruit and tea stains – pour it through the taut fabric from a height.
- Glycerine from the chemist gets grass stains out of fabrics.
- Use an oxygen-based bleach – or sunshine, which really does brighten whites, especially if you spread them on the lawn or a bush. But it also fades coloured clothing, so beware!
- Follow your grandma's example and boil tea towels and dishcloths for absolute freshness.

Dry cleaning

'Dry' cleaning means washing clothes in chemical solvents instead of water. This does avoid shrinking or felting, but the downside is that dry cleaning chemicals are toxic. They are especially dangerous to workers in the dry cleaning industry and are also harmful to the environment.

- Whenever possible, buy clothes that can be washed at home.
- Use a good cloth brush on wool suits – make strong sweeping strokes, against the nap and then with the nap – and air them after every wearing.
- Some garments labelled 'Dry clean only' – such as silk blouses – can be washed by hand.
- Look online for a laundry that offers 'wet cleaning'. This is a specialized service that does not require solvents.

Ironing

- If you use a tumble-drier, simply smooth clothes while they're warm to eliminate ironing.
- Do not tumble or iron fabrics until they are bone dry – they will last longer with gentle care.
- Some fabrics drip-dry beautifully. Linen blouses can be smoothed when damp – shape the collar with your fingers.
- Iron in large batches at a time to save electricity.
- Use traditional powdered starch for a crisp finish.

Moth protection

Avoid using chemical moth repellents with a few simple steps.

- Clean clothes before storage.
- Press with a steam iron or tumble-dry.
- Store sweaters in a sealed plastic bag that you can reuse each year. Freezing kills moth eggs, too, so rotate the bags through the freezer for a day or two before storing your clothes.

CHAPTER TWO

ORGANIC GARDEN, HEALTHY WORLD

Our bodies and our minds need fresh air, sunlight, seasonal temperature changes and the quiet of the natural world. Unplug electrical equipment and turn off your mobile phone. Open the windows and listen – really listen – to the birdsong and the breeze rustling the leaves. Lie on the grass and feel the earth beneath you. Relish the touch of raindrops on your face. Go outside on a hot summer's afternoon and let your body respond to the warmth. Don't fight the weather – enjoy it!

We can't truly care for the world if we obsessively isolate ourselves from it, and only by getting to know our own place on earth can we find the passion necessary to protect it. This makes gardening a vital part of going green, even if your garden consists of nothing more than two window boxes and a gritty scrap of weedy grass.

BACK TO BASICS

Here's the place to find ideas for going organic in the garden and using fewer plastics and fossil fuels in the process. Remember, energy in the garden should come from the sun and your strong arms! But if you're totally new to gardening, here are two little projects to show the wonderful vigour of plant life.

1. Cut an extra-thick slice from the top of five or six carrots, a good 25mm or so of the orange root and the green leaves. Place them cut side down in a saucer, on a little gravel or sand if you have some handy, and add 12mm water. Put the saucer on a window sill, keep the water high enough so the cut sides never dry out, and wait. Soon you'll see delicate fernlike leaves begin to unfold. I love this because the leaves are pretty – carrots are related to Queen Anne's Lace and other garden flowers – and because it's a great proof of the life force that resides in even a tired bag of supermarket vegetables.

2. This project will help you visualize what takes place beneath the soil when you actually get outside. And you'll end up with a tasty, nutritious salad! You can sprout many different seeds, but I recommend ordinary green lentils. Soak a small cupful in cool water overnight, then drain in a sieve. Either keep them in the sieve or in a special sprouting jar, and rinse them with cool water twice a day. Cover with a cloth or keep them in a dark place and watch what happens! A tiny pale root tip will appear

on the first day. When the roots are as long as the lentil is across, rinse them again, toss with vinaigrette and add finely-chopped onion and celery. Raisins or cubes of cheese are nice, too. Lunch is served!

Another fun way to start gardening is by growing bulbs that bloom during the winter, or even for Christmas if you plant at the right time. They can be planted in bulb jars that submerge just the bottom of the bulb, but grow better in commercial potting mix or a special bulb compost. After they flower, find a spot for them in the garden. Hyacinths produce huge heavy stalks of blossoms the first year, but the spikes gradually get smaller until eventually they look like large bluebells. You'll need to buy new bulbs each year for indoor potting.

CREATING AN ORGANIC GARDEN

You can't do a great deal of gardening from an armchair, but there's nothing more relaxing than curling up with seed catalogues and a note-pad to make plans for a new and greener garden. There are so many ways to garden, from planting a few fragrant herbs in a window box to growing great swathes of wild flower lawn – and everything in between.

Gardening is one of the best relaxants around because it gets us to tune in to the rhythms of the world. Growing things is all about nurturing. It's creative, like water-colour painting in three dimensions, and it calms your spirit and focuses your mind.

Delicious vegetables

With all the worries today about commercially-produced foods, it's a pleasure to eat what you've grown yourself. Growing food is a fundamental human activity. It provides an immediate connection with the vitality of nature and helps us to tune in to the signals of the natural world. There's nothing like picking greens for a salad fresh from the garden, or plucking a few leaves of oregano and parsley to toss into a sauce. And you really haven't lived until you've picked greens or dug carrots after the first snowfall!

- Broad bean (*Vicia faba*) plants are spiky, not beanstalk-ish, and have gorgeous purple flowers.
- Cauliflower comes in several colours: the purple varieties turn bright green when you cook them – but try them raw, too – and there's a lime-green cultivar with a swirly shape to the top.
- Red-flowered scarlet runner beans are both decorative and delicious; pick them when they're small and tender.
- Globe artichokes grow on huge thistle-like plants and are easy to grow in a border.
- Tuck herbs in among your flowers, in garden beds, or in window boxes. Try to keep some parsley, mint, rosemary and other favourite culinary herbs near the kitchen.

Fruit

New gardeners rarely think of growing fruit, but there is something breathtaking – awe-inspiring, even – about picking your own tender rhubarb stems or blackcurrants and they are so easy to grow. Most

fruits are perennials, so the easiest thing is to set aside a little space for a fruit bed and add to it gradually. Think about growing fruit trees, too. They can be messy, and eventually produce large crops, so you'll want to be sure that you – and maybe your neighbours – really like whatever it is you're going to grow.

Be creative. You can use raspberry bushes to create a hedge, or plant willow trees so you can try basket making one day. Quince trees are decorative, with bright pink or tangerine-coloured flowers and have attractive sulphur-yellow fruit that is excellent in pies and jam. This approach to landscaping, making the most of the productivity of your garden by using food crops in place of ornamentals, is known in green circles as 'edible landscaping', or, if it's confined to perennial crops, 'permaculture'. But for anyone who loves to eat and who hates to pay high prices for sad-looking supermarket produce, it's just good sense.

FEEDING THE SOIL

You may have studied ecology in school, but it takes a garden, a place you get to know through the seasons and year after year, to teach you why environmentalists pay so much attention to ecosystems. When heavy rain washes most of the topsoil from a bed that you carefully prepared for six new roses, reports of global soil erosion really mean something. If you manage to stop the run-off by planting a native wetlands shrub where the flooding used to start, you'll understand a lot about the kind of fieldwork ecologists are doing to protect and restore our environment.

Soil testing

If you are renovating a garden that has had no attention in a long while or you are breaking new ground, it's a good idea to buy a soil test kit. You can find them in hardware shops or garden centres. It's fun to play chemist and discover your soil's pH, nitrogen, phosphate and potassium levels. Once you know these, you know which plants will thrive and what you may need to add to the soil.

Magic mulch

Now that global warming is creating unpredictable weather patterns, the simplest way to feed the soil organically is 'mulching', which means putting a thick layer of organic material, i.e. a once-living material, on top of the soil, around your plants. Worms and other organisms in the soil break down, or decompose, organic materials into nutrients that plants can use. The mulch should be 7-10cm thick – you shouldn't be able to see the soil. This stops sunlight stimulating any weeds there may be around into growth.

Mulch is quite magical. It adds nutrients to the soil – a good organic mulch means you never need to use fertilisers – prevents weed growth and, most important in the dry days of summer, keeps moisture in the soil and drastically reduces the need to water. You can use almost any loose, dry organic material. Buy shredded bark, woodchips, well-rotted manure or straw; gather leaves, grass clippings and pine needles; or make garden compost. Bark and straw draw nitrogen from the soil in order to decompose, so if you use them it's a good idea to give plants a natural high-nitrogen fertiliser such as blood fish and bone.

A mulch can also effectively rid land of nasty perennial weeds. Put down a very thick mulch of newspapers, cardboard or even an old carpet to completely block out the light. It may take a year, but all but the most diabolical of plants will give up the ghost.

Nature's perfect recycling

Every organic gardener also likes to make his or her own compost. Gardens can use every bit you can make – and more. By composting kitchen scraps as well as weeds and other garden waste, you'll enrich your soil naturally. Fully a quarter of household waste can be composted in the garden. Some local councils now collect compostable waste separately. There are also schemes that allow residents who do not have the room to compost at home to bring their organic waste to a central depot and collect garden compost later. Even if you have no garden, composting will save precious landfill space and drastically reduce pollution from methane gases. It also saves money for everyone.

A composting bin of some kind is generally used. If your local council provides them at a discount, take your choice. You can buy one at a garden centre, too, or can build your own. The simplest way is to wire together old wooden pallets at the corners.

If you cook meals from scratch, you have quite a bit of kitchen waste. A well-sealed solar bin will make fast work of it, without your having to worry about flies or other pests. Add meat scraps only if you have a sealed bin and lots of material, so it composts quickly. Even the toughest peel – melon rinds, say, or grapefruit peel – will compost just fine, albeit a little more slowly.

The one critical thing is to add get oxygen in the pile, just as you want oxygen when you're exercising. This is easy: keep a pile of plant clippings and leaves next to the compost bin and add a layer of this loose, light material to every layer of kitchen scraps.

The pile should be kept damp – like a squeezed-out sponge – but not soggy. It'll soon be full of worms, working their magic. You don't need to buy special activators if you include some high-nitrogen ingredients – manure, seaweed if available, bone meal, urine (preferably a man's) or plants such as stinging nettles.

Nettles make a rich liquid fertilizer. If you have them available, put on gardening gloves, pull up young stems and pack them into a bucket or tub. Fill the bucket with water and cover it. Stir every couple of days and after two weeks, you can strain off a potent – and smelly – liquid fertilizer for all your plants. This can then be diluted and sprayed on to leaves to give plants an instant boost, particularly in cold, wet springs when nitrogen supplies are often low. Comfrey, treated in the same way, also yields a good fertilizer.

For peat's sake!

It's surprising how often we do things that are bad for the planet while thinking we're doing something wholesome and good. What could be better for a garden than natural peat moss? Some gardeners are still shocked to find out that peat is a limited natural resource not easily replaced. Almost 90 per cent of Britain's ancient peat bogs have been lost over the past century.

Home gardeners dramatically increased the amount of peat that they used during the previous decade and still use far too

much. There are, though, excellent alternatives to peat when it comes to improving your soil – garden compost; coir (milled coconut shell fibre) and leaf mould all do the job and come from sustainable sources. Make a point of seeking out these products, and asking for them, and refrain from buying peat and peat-based potting mixtures.

Leaf mould is richer in minerals than peat moss, and can be a far superior product, particularly for plants such as roses. It's easy to make your own: pile or bag autumn leaves, add a little moisture (but pierce the bag to allow drainage) and wait a year or two – or more – for them to break down. You can speed things up by chopping the leaves in a shredder or by running over them in several different directions with a lawn mower. This is different from composting. It is more dependent on fungi than bacteria and does not require the same high levels of nitrogen-containing materials.

WINDOW BOXES/PATIOS

You don't need rolling acres to make a start on the good life. Plenty of city dwellers bring a piece of the country to downtown Manchester or London's Docklands by taking advantage of window sills, balconies and even small slips of pavement where they can tuck a pot. There truly is a plant for every environment – think of those strange but appealing little 'air' plants that look like something from a sci-fi movie – so don't give up.

And you don't have to spend a lot on expensive pots, either. If you can poke around at a flea market, car boot sale, or in your parents' attic, you can find some interesting pots. You can even go for a Mediterranean effect by using old olive oil cans, although this isn't a good idea in a damp location. In a dry, sunny spot, old cans filled with bright zinnias and morning glories look terrific.

Just make sure your container has a hole in the bottom for drainage and is fairly large – at least 15cm in diameter. Bigger is better because tiny pots dry out too quickly. Who needs to suffer guilt pangs over a plant that's died of neglect? Fill the pots with top-quality potting compost so they get off to a good start and remember that plants are happier together, so put together a group. Inside, you can stand them together in a large tray on some gravel. It's always nice to grow something you can eat, even if it's just a few chives or some parsley. People are so impressed with a last-minute garnish of herbs you grew yourself! There's no reason not to mix different types of plants, as long as they need the same conditions. Chives like a rich moist soil – and so do primroses.

GARDENING TIPS

- Don't set out to garden alone: it's much more fun with other people, and when you're getting started you'll need advice and perhaps some morale boosting on the day you first spot lily beetles on your prize fritillarias.

- Share an allotment with a friend. You can cover each other over holidays and provide moral support during bad weather. The closer your allotment is to your home, the more often you will visit it.

- Don't overdo it. Gardening can be strenuous and, at its best, helps you stay fit. But it can be a strain if you're not used to the exercise. Learn to move correctly: when you dig, bend your knees and not your back; turn to throw that spadeful of soil only after you have straightened up; and make sure your tool handles are the right length for your back.

- Pay someone else to do hard physical work if the garden needs it or share the labour with a friend whom you can help out later in some other way – or whom you can pay in vegetables and flowers during the growing season.

- Always wear a hat and/or sunscreen in the sun.

- Buy or borrow the best tools you can afford – they will make physical labour a pleasure. A few basic tools of excellent quality are essential. If they are beyond your means, try sharing the purchase with a neighbour.

- Don't overspend – new gardeners tend to go a little crazy with excitement when they open a catalogue or walk into a nursery. Keep it simple the first year or two and you will enjoy gardening more, rather than feeling like it's another frantic competition or expensive hobby.

- Watch TV gardening programmes and borrow books from the library. Many newspaper gardening columns have excellent seasonal advice on organic gardening.

- Visit public gardens and go on garden tours whenever you can, and always take advantage of any opportunity you get to talk to the gardeners themselves.

GARDEN PLANNING

Don't undo the good that gardening does for the environment by driving back and forth to the garden centre every weekend. A little planning at the beginning of the season lets you avoid stressful and polluting driving. Call on your imagination and creativity: grab some coloured pencils and an attractive notebook and sketch what you'd like to see in a bed or even a pot on the balcony. Get on to the Internet and explore all the gardening websites; borrow some books from the library and, gradually, build your personal collection of gardening books and seed catalogues.

Think of your garden, large or small, as an outdoor room or as a series of rooms. Breaking up a garden with hedges and walls makes it seem larger and creates pleasant spots and hideaways at different times of day. Make sure there are places to sit – a large stone or sawn log can make a good casual seat – and create privacy with climbing plants on screens and trellises. Over time, you can add tables, chairs, play equipment, sheds and even shelves or other accessible storage.

Think of the neighbours before you plant a fast-growing hedge or position your new compost heap. If you have young children, you might even want to share garden space with neighbours – some people do this with great success and it's a much better story than lawsuits about out-of-control hedges. Incidentally, while hedges of traditional beech or yew are unlikely to alienate neighbours, all hedges need to be trimmed. If you don't want to make that commitment, a simple, elegant fence is a better choice.

Paving is a great way to create usable space – especially for a table and chairs. But don't seal off the soil with concrete! Use local stone or plain brick set in sand, without mortar, so water can run through. Low-growing plants such as creeping thyme are ideal to grow in the cracks: the tiny pieces soon grow into soft tufts.

Plant some white flowers near places you're likely to walk or sit in the evening, because they show up beautifully in moonlight. An oil lamp or citronella candles will provide gentle light.

And do remember to think of scent as well as colour when you plant. Place plants with fragrant leaves, including herbs such as the various mints, lemon balm and rosemary, in spots where you can easily pick a leaf, or where guests will brush against them. There are hundreds of varieties of sweet-scented garden flower, including many cottage garden favourites such as night-scented stock, wallflowers, clove pinks and sweet william.

Put out a light, switch on a star

Going out on a summer's night to watch an eclipse or a meteor shower is one of the best ways to connect with the beauty and mystery of the universe. If you live in or near a city, though, you've lost the view that everyone on Earth used to share, because of light pollution. The Royal Observatory at Greenwich has been forced to move twice because of light pollution – first to Sussex, then to the Canary Islands.

In the garden, use automatic timers to light fixtures only when they are needed. Remember to direct the beams downward and install energy-efficient, low-pressure sodium (LPS) lights.

Personal essentials

- High SPF sunblock (at least factor 15) applied 15 minutes before you go outside.
- A wide-brimmed hat – look charming as you garden.
- Waterproof slip-on shoes or clogs, so you can go in and out of the house with ease.
- Gardening gloves – it's nice to feel the soil with your fingers, but gloves prevent blisters when you do heavy digging, and if you use a hand lotion before you put them on, they will help to keep your hands soft and smooth.

CHOOSING PLANTS AND SEEDS

Believe it or not, even old plant varieties can be patented and companies can collect royalties on seed sales! Laws in the European Union have virtually eliminated many old varieties and cultivars by establishing a list of 'approved' seeds. This means that huge agro-chemical companies are often profiting at the expense of small, family-run seed companies and precious genetic variety is being lost. But around the world many people are fighting to preserve and revive heritage plants. You can help by buying plants and seeds from small firms that specialize in organic, open-pollinated, heirloom varieties. Join seed-saving organizations such as the Henry Doubleday Research Association in order to grow some endangered plants and learn to save their seed yourself.

TIPS FOR EASY GARDENING

Put in vegetables and herbs near the kitchen to make it more convenient to pick them for meals.

Grow things you love to eat. There are no rules about where to start! If you happen to want ruby-coloured heads of radicchio in your garden, grow that.

Mix flowers and vegetables and herbs and fruit for a truly modern garden. Many vegetables are beautiful in a border. It's easy to throw in flower seeds when you're planting to brighten the vegetable rows.

Plant bulbs in the autumn and don't forget the really early ones such as snowdrops and winter aconite – what a lift to see flowers bloom during the final days of winter!

Choose pest-resistant varieties adapted to your climate and garden conditions.

Plant perennials – plants that last for many years. Their initial cost may be high, so add them gradually.

Plant self-sowing annuals. It's fun having seedlings turn up in odd corners. Poppies, forget-me-nots and many others will re-seed themselves year after year.

Weeds gone wild

A new problem for gardeners – and a serious global ecological challenge – are plants labelled non-native invasive species. These are plants (and animals) that have been introduced to a new area and have spread out of control. Japanese knotweed, which has spread rampantly along many British rivers, is a good example. These plants were introduced innocently in the 19th century by gardeners who loved to experiment with beautiful, exotic plants. They are easy to grow and spread quickly because they have no predators or competing plants here to keep them under control, as they do in Japan. They drive out native species that insects and animals rely on for food and upset the balance of nature at an astonishing speed. In fact, non-native invaders are second only to habitat destruction and decline as threats to biodiversity.

You'll probably notice that seed and plant catalogues are including more information about this issue. Plant sellers and plant collectors are both becoming more conscious of the risks of their hobby. You can help by always choosing native and traditional varieties. You should never indiscriminately move any plant into or out of the wild.

BIODIVERSITY IN THE BACK GARDEN

One of the major effects human beings have had on the earth has been to contribute to the loss of species. This isn't simply because we have hunted a species to extinction – as we did the passenger pigeon or some species of whale – but because our activities change the environments (or ecosystems) in which they live. Even a single new house disrupts water flows and migratory patterns amongst the tiny creatures that live in a wood. This is one reason conservationists push so hard to limit encroachment on undeveloped land and to encourage town planners and developers to focus on areas that have already been changed.

Your gardening efforts will be a positive force if you think first of what must once have grown there. Native trees and wild flowers are easier to maintain and require less water and fertilizer than many standard garden varieties. If you live somewhere that was once woodland, for example, plant wood anemone, lily-of-the-valley, primrose and bluebells. Similarly, if you live close to the coast, try planting wild flowers such as thrift, campion and sea holly. It always makes more sense ecologically to fit your plants to the environment, rather than the other way around.

You'll find loads of advice online, with lists of plants for your particular area. The mainstream gardening publications and broadcasters are well aware of environmental issues and provide a wide range of helpful resources to home gardeners.

The loss of wetlands is a global problem, because these areas create an essential buffer for changing water levels and are a vital habitat for many bird species. So if you are privileged to have a boggy site, rejoice! You can create a stunning bog garden full of precious native plants and simultaneously protect one of the planet's most delicate ecosystems. Growing a good mix of plant species – especially native and old-fashioned hardy cottage garden plants – and planting open-pollinated, or non-hybridized, seeds is another way to support biodiversity and create an environment that is attractive and easy to maintain. As a bonus, biodiversity itself helps to protect your garden from being plagued by insect pests and plant diseases.

ATTRACTING WILDLIFE

A pesticide and chemical-free garden is a welcoming home for animals whose natural habitats are being chewed up by new roads. Many of these creatures lend their help to your organic gardening efforts. Toads, for example, gobble up slugs and snails, while most birds are beneficial, too. They can eat their weight and more in insects every day!

- Plant native trees and shrubs rather than exotics. Some of the most famous gardens are based on native plantings. Not only do natives help to create a healthy garden ecosystem, they also require less watering and other attention.
- Choose native perennial and self-seeding flowers and herbaceous plants. Some nurseries specialize in these plants and can advise on what's right for your area.
- Plant old-fashioned scented cottage flowers to attract bees and butterflies. There's nothing as wonderful to bring into the house as a bouquet of fragrant sweet peas or one of old-fashioned roses.
- Turn part of your lawn into a wild flower meadow that you cut only twice a year. This saves labour and makes a striking

display. For example, cover a steep bank with black-eyed Susans, ox-eye daisies and fancy goldenrod cultivars.

- Plant buddleias, scarlet lobelia and others to feed butterflies.
- Don't buy 'double' flowers; insects find it difficult to find the nectar in among the profusion of petals
- Don't make things too tidy: the odd fallen log and pile of pots or stones can be attractive and also create a shelter for helpful garden creatures. Of course, it's best to avoid offering accommodation to larger pests, so bear this in mind.
- Leave the weeds – or at least, a few. Set aside a little space for nettles for example. They are edible and nutritious when small and sustain the caterpillars of several butterflies. Remember, a weed is just a plant in the wrong place, Cultivars of buddleia, a common city weed in Britain, are prized garden plants in the US, while native American goldenrod and sumac, considered weeds in the US, are specimen plants in U.K. gardens.
- Put in a small pond. Frogs urgently need them to live and breed in. Frogs make wonderful garden co-habitants because they munch on insects day and night. You may wake up one Spring morning to a pond that is thrumming with tiny frogs!

ECO-FRIENDLY PEST CONTROL

Strong plants tend to be disease-resistant and organic gardeners avoid chemicals by making use of sensible prevention, sophisticated planting techniques and natural pest control products. The most important element to pest control is to plant a wide variety of native species. This ensures that no pest can run riot, because there will be a limited number of host plants for it.

Another important organic approach is to encourage friendly insect predators such as ladybirds and hoverflies, both of which are far more effective against aphids than chemical sprays. The more wildlife you can encourage into your garden, the more pests will be picked off your plants. Birds in the garden – sparrows, blackbirds, thrushes, tits and robins – account for several invertebrate pests, while frogs, newts, toads, hedgehogs and ground beetles help with the slugs. Do not be too excessive about keeping the garden tidy. You want to provide some habitat for these friendly creatures.

Weeds can be controlled by close planting, mulches and hand weeding. Let pulled weeds dry for a few days to ensure they don't come to life again – this sounds like sci-fi, but you'd be surprised how determined some can be – then put them in the compost or, in the case of weeds such as ground elder or bindweed that grow from a fragment of root, bin them. Local by-laws make it illegal to burn garden debris in many areas, but even if it weren't, it is a waste of nutrients. Use layers of cardboard, carpets or black plastic to smother tough perennial weeds and clear ground for planting.

A few well-known plant-derived pesticides, such as sabadilla, pyrethrum and neem are sometimes seen as suitable for organic gardens. A variety of pest control products available by mail order and in garden centres can be supplemented with mechanical controls such as sticky traps and physical controls such as picking off beetles or squashing aphids. It's also wise to select cultivars that are both pest- and disease-resistant when possible. It's easy to make your own insect repellent sprays. Simply use a blender to mix water and a combination of garlic, dried or fresh hot pepper and, if you have them, marigold flowers. Strain the mixture through a very fine sieve and spray it on plants to repel aphids and other insects. You should spray in low light to avoid injuring the leaves. Experiment to learn what percentages of each ingredient work best on various plants – and don't rub your eyes!

Baking soda can eliminate powdery mildew and some other fungal diseases. Mix a teaspoon (5 ml) of baking soda with one and a quarter litres of water and add a squeeze of dishwashing soap – not detergent – to help it stick. Spray it on leaves early in the morning, before the sun is bright, at least once a week and repeat after every rain.

Hand-picking insects can become a game for children. Let them knock bright scarlet lily beetles, for example, into a tub of soapy water. Some people swear by soap – not detergent – spray and you can make double use of wash-water if you happen to have roses growing outside the door. Just toss the soapy water over any plant that suffers from aphids.

Organic gardeners also use physical barriers such as small rings of cardboard to protect seedlings from cutworms. Floating row covers and floating insect barrier covers prevent many insects from laying their eggs near or on the plants and can effectively protect almost all young seedlings.

Snails and slugs, which are extremely destructive in mild climates, will not cross sharp sand or wood ash – provided it stays dry – and they can be attracted by, and indeed drown in, saucers of beer. Put orange or grapefruit peel on the ground: slugs enjoy the damp shelter under them and can be scooped up and disposed of *en masse*. Alternatively, go out with a torch one evening after rain and pick them off your precious plants. Collect them in a lidded bucket, and empty it out in waste ground 100 metres or more away – slugs and snails can and do return to their old territories. Chemical slug pellets don't belong in a green garden. They can kill friendly insects and are dangerous to all wildlife, especially the birds that swoop down to eat the dead slugs.

Happy companions

Companion planting is the practice of planting different plants in close proximity to each other because they either contribute useful compounds to the other or because one repels an insect or animal that preys on the other. Native Americans were doing this when they grew the 'three sisters' – beans, corn and squash – together. As a general rule, try planting strongly scented plants such as mint, marigold, garlic or onions next to any plant that suffers from insect damage, but pay attention to rules such as keeping onions and peas and beans far away from each other.

Common combinations to try

Carrots and tomatoes. Candytuft (*Ibericum*), dill, sage, or mint will all help to deter cabbage moths from cabbages, broccoli and other brassicas. Nasturtiums can be planted to deter aphids and pests that prey on cucumbers, pumpkins, courgettes and squashes.

Insect repellents

There are seasons when insects seem to spoil the great outdoors, but many repellents are ferocious poisons. Rather than apply commercial repellents, which contain Deet (diethyl toluamide), a strong irritant which can eat through plastic and dissolve paint, try the following:

- Rub your skin with vinegar that you've dropped on to some cotton wool. The smell disappears as it dries, but makes you taste nasty to passing insects.
- Rub oil of citronella or pennyroyal, diluted in a little vegetable oil, on to your skin.
- Put up a bat box. Bats are endangered and need lots of new places to roost – and they can eat thousands of mosquitoes and other flying pests in a single night.
- Take a careful look for any standing water – even a piece of broken clay pot can be a breeding ground for mosquitoes. Make sure water butts are covered.
- Don't apply repellents until you are sure they are necessary – unlike hats and sunblock, these products should never become routine.
- Learn to live with the occasional fly or mosquito. When insects are really fierce, stay out of their way.
- Read labels carefully – I've seen a typical chemical product labelled 'eco-spray'.
- Look for new products such as 'bug-buttons' that keep chemical repellents off your skin.

Four-legged pests

Gardening for wildlife can bring in not so welcome animals. Small rodents such as rabbits, mice and squirrels can damage plants by eating leaves, seeds and bulbs and gnawing bark, but by far the most dangerous is the rat. In the town or the country, rats are never far away. If you see them in your garden in the daytime, call pest control. Weil's disease, carried by rats, can kill. It is spread in rat's urine and can contaminate standing water. Don't put your hands in garden ponds if you have nicks or cuts on them.

WATER CONSERVATION

Apply thick mulches to keep water from evaporating from the soil. Mulches make a fine cover for earthworms and improve soil quality in general.

- Water early in the morning or at dusk so the water soaks in rather than just evaporating in the sunshine.
- Use soaker hoses – which are porous or full of tiny holes – rather than sprinklers. Less frequent but deep watering encourages plants to grow deep roots.
- Save rainwater in a barrel or water butt and use it to water the garden – or wash the car. Cover it to keep out debris and also prevent mosquitoes from breeding in it.
- If you live in a hot, dry climate, look up Xeriscape on the Net. It's a method of planting with native and similar species that need almost no supplementary water, even in the driest

weather. This is a great time-saver. Native plants are ideal because they attract butterflies and create an environment for wildlife and they're also easy to maintain.

THE ECO-FRIENDLY GARDEN SHED

Not only do we want to minimize driving to the garden centre, but also to use long-lasting equipment and tools. There's a lot of junk out there, designed to attract the excited novice. As ever, the most important plastic to avoid is the stuff designed for the trash – for immediate disposal. Buying things made out of plastic that last doesn't give the same problems.

- Turn up your nose at garish hanging baskets and elaborate trellises. It's easy to make lovely trellises and arbours out of simple branches tied with cord or wire.
- Instead of buying peat pots at the nursery, reuse pots or cell packs and fill them with your homemade compost sifted with sand or a basic organic potting mix.
- Raise your own seedlings and share extras with friends.
- Choose biodegradable, untreated gardening string or use torn strips of cloth to tie up your tomatoes. After harvesting, both plants and ties can go straight on the compost pile.
- Learn to propagate plants. This is an amazing way to save money, too. The key is to know whether a plant is going to re-seed itself, such as hollyhocks, columbines, or love-in-a-mist or will have to be divided, such as astilbes, hostas and irises.

- Choose garden furniture made from sustainably-grown wood rather than endangered tropical teak. Furniture made from metal and plastic can be a good choice, too, because it often lasts longer than rattan pieces.

TRY THIS

Yoghurt tubs and plastic packing trays are useful for planting seeds. You can also use egg cartons. Cut the bottom off plastic gallon water bottles to make mini-cloches – a greenhouse for individual plants – but take off the cap and pay attention to the temperatures inside them when the sun is shining to keep them from baking your plants.

TIPS FOR AN ORGANIC LAWN

Spike with a fork or with the spiked boots available from gardening shops to aerate the soil, making sure you penetrate any thatch that may have accumulated.

Allow extra lawn space to grow tall and wild, and plant wild flowers to create a lovely meadow.

Choose a hard-wearing seed mixture appropriate to your area. It'll be easier to maintain and the weeds won't be so obvious.

✳   Or try another kind of ground cover: a clover lawn can be beautiful and hardy and in some areas you can grow other plants, such as camomile, as a lawn.

✳   *Set your mower blades as high as they can go. Lawns do best if cut high and often, and weeds find it hard to gain a foothold – they won't get enough light to sprout. Leave clippings to replenish the soil, or add to compost.*

✳   Use a push mower, especially for small lawns. If you prefer a power mower, choose an energy-efficient model that will also mulch leaves.

✳   *For healthier grass, make sure mower blades are sharp and water infrequently but thoroughly, early in the morning to avoid loss through evaporation.*

✳   Use an old kitchen knife to dig out weeds with deep taproots such as dandelions and docks. New dandelion leaves make a wonderful springtime dish, like escarole, so the ideal time to clean up your lawn is late afternoon, when you can bring in your basket of greens to fry with garlic in some good olive oil.

CHAPTER THREE

BON APPETIT!
Better food, safer planet

Eating is powerful. Food is fundamental! Much of our history has been shaped by the availability of the foods we eat, and much of our effect on the natural world has been shaped by our growing, storing and transporting those foods. But our relationship with food has changed dramatically during the last century or so. Now, half the Earth's six billion people live in cities and do not grow their own food. In the most highly developed nation, the United States, five per cent of the population grows food for the other 95 per cent, creating an elaborate and expensive food distribution system.

While people in poor nations lack clean water and basic nutrients, others in wealthy countries spend time and money trying to compensate for excessive eating. It's ironic that though we have plenty to eat – more than enough – food has become an aspect of life that's fraught with anxiety, and even fear.

TIPS FOR YOUR HEALTH

* *Eat and drink foods that have been grown without chemical fertilizers and pesticides.*

* Be adventurous – eat a wide variety of foods and emphasize locally grown products.

* *Stay low on the food chain: eat fruits, vegetables, and whole grains rather than animal products.*

FAST FOODS, SLOW FOODS

We worry about salmonella and pesticides, about mad cow disease and carbohydrates, not to mention the question of what genetic modification is going to do to us. Many of these dangers are caused by globalized, consolidated agribusiness and by food production and processing systems that are driven solely by profit. If we wish to create a peaceful world, it's imperative that we develop healthy, equitable, and sustainable farming and distribution systems. Changes that improve the environment also improve your health and make you feel more confident about the food you serve, so it's a pleasure to eat with family and friends.

Around the world, traditional cuisines developed in ordinary kitchens where people used the most abundant and least expensive local foods. But now people routinely eat meals away from home, at cafes and snack bars, restaurants and pubs. This is a mixed blessing in environmental terms. Eating in restaurants can be good for the environment because foods are purchased in bulk, saving packaging, and are prepared in large quantities, thereby saving energy. But sadly, even expensive restaurants serve processed and prepackaged foods.

Environmentalists and food and labour activists criticize fast-food restaurants on a number of fronts: for contributing to the destruction of tropical rain forests and the loss of biological diversity, for the large amounts of paper and plastic waste that they add to the disposal stream, for the unhealthy food that they serve, and for their unfair labour practices. But the fact is that time is at a premium for most of us today and that fast foods are just that – fast – and cheap, too. We need genuine alternatives that are fast *and* healthful. Traditional fast foods are *real* street foods that local people make themselves and sell to earn a living. They use local, seasonal ingredients and have great character.

Look at cooking as a pleasure, not as a chore! If we can focus on the conviviality of eating instead of on the 'convenience' of fast food that allows us to rush off to check our messages, we'll refocus our lifestyle and improve our relationships, too.

Message it! Contact Slow Food at http://www.slowfood.com *or* international@slowfood.com (an international organization that "champions the firm defence of quiet material pleasure as the only way to oppose the universal folly of the Fast Life.")

TIPS FOR COOKING AT HOME

* *Buy in bulk. This practice not only allows you to buy less packaging, but also means you'll rarely run out of vital ingredients.*

* Be ambitious. Learn to make a cheese soufflé or profiteroles – in fact, both are surprisingly fun and easy.

* *Simplify. You don't have to peel potatoes, and why not use cookware you can take to the table?*

* Stock your pantry with staples that last and with lovely fresh foods that don't require cooking: cheese, good bread, fruit, vegetable sticks, yogurt.

* *Offer to help a skilled friend so that you can watch and learn more about cooking.*

* Make a list of foods you want to be able to prepare yourself and practice preparing one every week.

* *Save energy for the planet – and your own energy too – by doubling or even tripling recipes and freezing the extra to eat later.*

Pots and pans

Pay more and get the best. The safest materials are stainless steel, cast iron, glass, and enamelled cast iron. These conduct heat so evenly that food rarely sticks or burns. The lids seal correctly and the cookware is a pleasure to use. Stay away from aluminium pots as aluminium intake has been linked to a number of brain disorders. Acids in food dissolve it, bringing small but potentially injurious amounts into the food you're cooking. Also, don't buy pans with nonstick coatings. When they are overheated they release toxic fumes, including perfluorooctanoic acid (PFOA), which has been found in the blood of 96 per cent of children tested in 23 US states. PFOA is linked to human birth defects.

FRESH, SEASONAL FOOD

When I say 'local food', I mean food that has been grown and produced near my home – maybe even in my own garden. Eating local food has important environmental and health benefits. It's important to reconnect with your country's food heritage by eating food grown in local regions and in local soil.

In developed countries, the cost of processing, storing, and distributing food accounts for more than 50 per cent of the total food bill. Although processing is not necessarily bad – people have preserved and stored food for thousands of years – these costs reflect energy and resource uses that contribute to air and water pollution, global warming, and damage to ecosystems. Fortunately, it's becoming increasingly easy to source food locally.

- **Grow your own:** What's more local than your own garden?
- **Co-op and box schemes:** Look for a service that delivers organic food to your door on a weekly basis – encourage your provider to source foods from your region.
- **City farms:** These provide employment and a chance for city children to experience farm life and taste really fresh foods – you can often buy milk, vegetables, and other products from them.
- **Allotment schemes:** Local councils in urban areas rent out allotments for gardeners, although there may be waiting lists. Join like-minded allortmenteers in adopting organic practices.
- **Farm shops:** Large or small, these rural retail enterprises are good for farmers and make a fun stop on any car trip and a chance to connect with place.
- **Pick-Your-Own:** You do some of the labour and as a result, get absolutely fresh berries and other produce at great prices. Search for organic self-picking farms.
- **Farmers' or Green markets:** These are increasingly common in cities and rural areas, too. Selling directly to urban consumers helps farmers economically, and you benefit, as well as they bring the farm produce to you.

Natural food stores are the best place to look for locally-grown foods. Projects such as Berkshire Grown, where I live, are designed to get local produce and meat into ordinary supermarkets and restaurants. So choose locally grown whenever there's a choice – you benefit, and so does your region and the world!

TIPS ABOUT ECO-PACKAGING

Contrary to popular opinion, using a plastic carrier bag is not the worst of environmental sins – a cloth or string bag is the best choice, but don't lose sleep over this issue.

* *Carry your own shopping bags or use the store's boxes.*

* In Thailand, women traditionally wrap food in banana leaves. Encourage manufacturers to devise similar low-impact packaging materials.

* *Animals can become ensnared in plastic netting, and deer have been known to die as the result of swallowing plastic bags. The plastic rings that hold 6-packs of cans and bottles can be lethal, too.*

* Buy in bulk and in returnable/refillable containers.

* *Buy less canned and bottled food.*

* Choose concentrated products and dilute them.

* *Avoid packaging that consists of mixed materials, paper and/or plastic and/or foil, because they are too expensive to recycle – individual juice boxes are one example. Urge manufacturers to design for recycling or reuse.*

AU NATUREL: GOING ORGANIC

Eating organically grown food whenever you can is important to your health and the planet's. Studies have found more than 50 pesticide residues on conventionally grown produce – a potent chemical cocktail to ingest throughout our lives. Eating organic food is the healthy choice, especially for children. Organic agriculture uses biological pest control, crop rotations, mechanical weeding, and animal manure and plant wastes for fertilizer.

Agribusiness uses compounds such as oestrogen as a growth stimulant for livestock, and many pesticides contain chemicals that mimic oestrogen. And we wonder why sperm counts are declining and why there's an increased rate of breast and prostrate cancers. Another result of chemical methods of farming is nitrate contamination. Nitrate contamination causes crop damage, forest die-back, ground-level ozone pollution, contaminated ground and surface waters, and damage to coastal fisheries from algal blooms. Plants need nitrogen, and non-organic agricultural systems depend

Did you know? Seventy per cent of organic food sold in the UK is imported, as domestic supplies are insufficient and consumer demand for such staples as tomatoes and apples, for example, takes no account of seasonal factors. Organic nurseries in the EU and further afield supply this demand. Popular tropical items such as chocolate, coffee, tea, bananas and many other fruits are routinely imported.

on high nitrate fertilizers. Chemical fertilizers are easily soluble, unlike the more complex, slow-releasing natural sources of nitrogen, and a high percentage of the nitrates in commercial fertilizers run off into our water systems. Organic farmers make use of animal wastes, returning them to the soil in a composted form rather than spreading them raw on farm fields or dumping them into our water systems.

Did you know? Recent studies show that highly-coloured foods – dark green, red and orange in particular – contain more vitamins, flavonoids and other beneficial nutrients.

The most important organic foods to buy

The following foods require high inputs of chemical fertilizers if they aren't organically grown, or require especially high levels of pesticides. So, go organic with these first:

- Corn
- Bananas
- Apples
- Any food containing animal fats (cheese, butter, milk)
- Peanuts and peanut butter
- Rice
- Strawberries
- Breakfast cereals

Agricultural pesticides are washed from fields and orchards by heavy rain and get into the groundwater. This means that meat, game, fish, grain and dairy products can all contain pesticide residues, but the greatest problems arise with otherwise extremely healthful fruits and vegetables. Around 70 per cent of all apples, carrots and lettuces sold in Britain in 2000 – and more than 80 per cent of pears – had some pesticides, many from chemicals designed to improve the look, not the health, of the product.

The best way to avoid these is to eat only organic produce, but if this is not possible, remember a few simple rules. The most important is to wash all fruit and veg thoroughly in cold water before it is peeled so the paring knife does not contaminate the flesh. Avoid soft fruits that cannot be scrubbed.

FOOD FOR THOUGHT — SAFETY

Factory farming has another dangerous downside – many chickens are contaminated with bacteria that can result in food poisoning. The numbers continue to rise, with a million cases of salmonella and the intestinal bacterium *E. coli 0157* reported annually. Contamination of food supplies is on the rise because of contemporary intensive rearing and slaughtering methods designed to produce cheap food. Global markets and distribution systems make it easier for new diseases to develop and spread, too.

The trucks and planes that move food supplies around the world are a major source of air pollution, especially because they refrigerate their food shipments to minimize bacterial growth and contamination. Although we, as consumers, have come to expect certain foods to be available all year long, whether in season or not, we can't always be sure that the countries of origin are using the procedures necessary to prevent food contamination.

Food industry safety experts focus on ways to destroy pathogens, from irradiating foods to treating them with chemicals or hydrostatic pressure systems. The ecological approach is to buy local food that has been raised and produced with care; to eat less meat and more fresh, unprocessed foods; and to make simple changes that can help to reduce food-borne illnesses.

Washing your hands, especially before you eat, is the most important step you can take to protect yourself from pathogens. Scrub under your fingernails too. Food safety experts recommend that you wash for at least 20 seconds: the water can be warm or cold. Ordinary hand soap is just as effective as antibacterial soap, and unlike them, it does not contribute to microbial immunity to common antibiotics.

As for your food preparation surface, a much publicized US study showed that, contrary to expectation, wooden chopping boards are safer than plastic. In the study, almost all the bacteria on wooden boards died, while in contrast, some bacteria actually multiplied on plastic boards. The solution is to use more than one chopping board: one for meat, one for bread, and another for vegetables and fruits – especially if they are to be eaten raw.

GENETICALLY MODIFIED FOODS

For thousands of years, people have been selecting crops based on their genetic traits, and this selective breeding has provided us with staple crops such as oats, wheat, and rice. But in recent years scientists have begun to modify the genetic traits of crops in the lab with astonishing (and alarming) results.

The food industry is focused on increasing corporate profits, so the emphasis has been on the genetic modification of crops so that it costs less to grow more. Inevitably, consumers have objected to having these foods imposed on them with inadequate testing and labelling. The right to choose what's in your food should be paramount – and many observers see the pressure to extend the use of GM foods as a blatant disregard for public opinion and consumers' right to know what they are eating.

Genetically modified (GM) crops are almost certain to be a part of our food future – the biotech industry is a huge global player – and they may also have some health and environmental benefits. But what's crucial, ecologically speaking, is that potential products be tested extensively and that consumers receive complete information about the test results of various products, as well as information that lets them know if any GM crop is included in a particular product.

In a recent study in the UK ten out of 25 soy products tested were tainted with genetically engineered material. Eight of the ten were labelled either as 'organic', which should indicate the absence of transgenic ingredients under Soil Association rules, or explicitly as 'GM-free'. And in the USA recent reports by the Union

of Concerned Scientists and the Academy of Sciences highlight the threat of GM crops contaminating conventional seeds and harming the environment as altered genes spread. But the US Department of Agriculture is considering allowing the spread of these altered genes, including those from pharmaceutical crops, in conventional seeds without reviewing their impact on human health and/or the environment.

We must protect the UK seed supply from contamination. Regulations should ensure that GM crops are monitored. Biopharm crops producing industrial chemicals and pharmaceutical drugs should not be planted in the open, nor should biopharm genetic material be engineered into food crops. As consumers, we should buy organic corn and soy products whenever we can and ask shop managers if such products are GM-free. Check out the Greenpeace GM shopper's guide on http://www.greenpeace.org.uk.

Did you know? GM soybeans contain 13 per cent fewer beneficial plant oestrogens than non-GM soybeans.

BIODIVERSITY ON YOUR TABLE

Biodiversity is diversity in the environment measured by the numbers of species of plants and animals it contains. Variety is the spice of life, and biological diversity is life itself. The more species of plants and animals our planet has, the healthier it is.

Earth, though, is losing biodiversity at a terrifying speed. The World Resources Institute has listed seven causes for this loss: our failure to realize the value of biodiversity; population growth and increases in resource consumption; poorly conceived policies; global trading systems; ignorance of species and ecosystems; inequity of resource distribution; and the interaction of these six factors.

Deforestation in tropical regions also contributes to the loss of biodiversity. Deforestation causes crop gene pools to shrink in a similar fashion to overexploitation of plant and animal species and pollution. Species also can become extinct through a combination of hybridization, disease, predation, and competition for food and nest sites. You can help slow the loss by becoming adventurous in your choice of foods, and eating the unusual. Select traditional varieties of tomatoes, potatoes and apples, and look for unfamiliar grains such as quinoa, kasha, spelt and millet.

Here are some more tips:
- Buy a wide variety of vegetables and fruits.
- Seek out and attend an apple tasting of local and old varieties such as Worcester Pearmain and Blenheim Orange apples.
- Buy meat from a farmer who raises rare old breeds.
- Buy multi-grain breads – they're delicious.
- In your garden, plant open-pollinated seeds and old roses.
- Make and serve traditional drinks such as cider, perry and home-made country wines.
- Eat wild crops such as nettles, dandelions – a delicious and nutritious salad – crab apple, bramble and elderberries.

Did you know? The UK once had more than 6,000 varieties of apples, with favourites in each region. Now supermarkets tend to sell half a dozen varieties, most grown abroad.

RED MEAT, YELLOW LIGHT

Almost every culture throughout human history has used animals for food. Traditional farming methods depend on the use of animal manures as fertilizer. 'Mixed farming' creates a neat ecological cycle, in that animals graze fallow land, eat grasses and seeds that would otherwise be wasted, and return nutrients to the soil in manures. Free-range chickens keep crops healthy by eating large numbers of insect pests, and provide high nitrogen manure to fertilize the next crop. Goats and sheep are crucial in many societies as they provide both meat and milk – as well as wool and leather – from land that is often unsuitable for farming.

The commercial need to produce meat cheaply is clearly the primary cause of bovine spongiform encephalopathy (BSE), or mad cow disease. Modern meat-rearing methods have also contributed to the rise of 'superbugs', bacteria that are highly resistant to antibiotics. Large-scale livestock operations routinely add antibiotics to animal's rations, not to fight disease but rather to fatten up the animals. This practice causes the animals to grow unnaturally quickly, and contributes to the growing problem of drug-resistant bacteria.

In addition, more than half the world's grain harvest, much of it grown in developing countries, is used to feed livestock. To produce a quantity of beef requires over 15 times its weight in grain, as well as a considerable amount of water. This concentration of protein is both expensive and inefficient, and concentrates pesticides that are found in the grain or other feed. Pork production is more efficient than beef. It requires six kilos of feed to produce a kilo of meat, while the protein conversion rate for broiler chickens and for eggs is about three to one. Such practices are an inexcusable waste of food resources, and most environmentalists prefer to promote a practice they call "eating low on the food chain".

Traditional cuisines in most parts of the world, including much of Europe, use meat primarily as a seasoning or as a special treat rather than the focus of a meal, and many cuisines that that use little or no meat are becoming more popular. This trend has helped many of us break away from the traditional British dinner of meat and two veg.

Veganism is a lifestyle choice that precludes eating any animal products at all. This choice is a demanding one, because vegans must eliminate great instant foods such as cheese and eggs. However, many wonderful vegan dishes can be prepared – consult almost any Asian cookbook for ideas.

As a group, vegetarians and vegans have lower blood pressure and cholesterol levels, substantially lower rates of heart disease and cancer, and tend to be slimmer than meat eaters. Some people claim that vegetarians and vegans don't get enough protein, and although this claim is true of many people in the Third World, in the West we get more than enough protein without eating meat or soy substitutes. Eating excessive amounts of protein has even been linked to premature aging and degenerative diseases.

Those of us who enjoy meat can now get it from animals that are raised without growth stimulants, hormones, and additives – the RSPCA's meat-monitoring scheme and the Freedom Food label ban antibiotic growth stimulants. Some suppliers sell meat that comes from rare old breeds, which means that you can promote biodiversity as you grill your pork chops. These meats, from fine old species such as Old Gloucester pigs, are expensive but delicious. If you are buying from the supermarket, lamb is probably a better bet than other commercially-reared meat as sheep are primarily range animals, and are not fed with grain.

Look for meats that are labelled as grown without additives such as hormones and antibiotics. When possible, buy meat from health food stores or from organic farms and co-operatives that distribute meat throughout the country. Check for suppliers of meat and other organic produce at http://www.alotoforganics.co.uk.

Living lightly: green and low-carb diets

Obesity is becoming an alarming social problem, rivalling cigarette smoking as the leading cause of death in the United States and afflicting an increasing number of children in Britain and throughout the world. We are overweight because we don't exercise enough and because we eat too much of the wrong foods. Consequently, making choices that will do the environment good – such as walking instead of driving and eating home-grown fruits and vegetables – will do our waistlines good, too!

But the trend in dieting is to eat low-carb food, which too often means eating lots of meat, fish, eggs, and cheese. If the billion obese people in the world went on an Atkins-style diet, 400,000 more square miles of land would have to be cleared and ploughed to grow enough soybeans and corn to feed enough additional animals to feed these dieters. Air and water pollution, overgrazing, and methane production – which contributes to global warming – would surge.

The fundamental problem with carbohydrates is that we eat them in highly refined forms. Kasha, brown rice, and whole-grain breads don't affect the body the same way that refined sugars, polished rice, and white flour do. It's much wiser to follow a moderate, healthful version of the South Beach low-carb diet, with plenty of vegetables, soy and beans alongside moderate amounts of meat, fish, and eggs. Check Indian and Middle Eastern cookbooks for delicious low-carb dishes that will help you control your weight without putting additional pressure on the planet.

SAVING THE SEVEN SEAS

Fish is good for us, but overfishing is taking a heavy toll on our oceans, lakes, and rivers. Most fisheries are being exploited, and many once-abundant sources of fish have essentially vanished because of fishing methods as well as consumption patterns. The result is long-term and even permanent damage to ecosystems.

Another result of overfishing is human hunger, because fish is a critical protein source for at least a billion people, most of them in Africa and Asia. Some experts recommend that we eat only farm-raised fish until global fishing is properly regulated. You should look for the Marine Stewardship Council fish certification logo and the Turtle-Safe Shrimp logo when you buy seafood. Always ask your fishmonger about the source of any fish you buy. You may know that studies have shown that farm-raised salmon often contains higher levels of PCB contamination than is usually considered safe – if you eat over 250 grams of it a month. However, if you vary the types of fish you eat, this shouldn't be a problem. It's also important to know whether a fish is endangered or not. Check www.greenpeace.org.au/oceans/ for up-to-date information about endangered fish species.

Some fish (especially swordfish, shark and king mackerel) are also contaminated with heavy metals such as mercury, in the form of industrial by-products released into the seas. Use the Internet to keep up to date about fish safety.

PLAY FAIR: BUY FAIR TRADE

Some of our favourite drinks and treats are grown in tropical countries where working conditions are bleak and contemporary agricultural systems put great pressure on ecosystems. Coffee and tea are staple drinks throughout most of the world. Although it's ecological – and delicious, too – to drink locally grown teas, most of us also want our capuccinos and Assam tea, so let's look at how to encourage sound agricultural practices and fair trade.

Fair trade is a global movement to ensure that the producers of the goods we buy get a fair price and that they have decent and safe working conditions.

Consider chocolate. Worldwide, chocolate, made from the bean of the cacao tree is big business. The United States, for example, consumed $12.4 billion (£7 billion) in chocolate in 2000 alone. But cocoa farmers are routinely underpaid. As a result, cocoa farmers in Central America and West Africa reduce costs by taking their children out of school to work on the cocoa farms or, worse, by using slave children from even poorer nations such as Mali and Burkina Faso. For every kilogram of chocolate produced, 180 cocoa pods must be picked and sliced open to remove their beans. Child workers have to use machetes for this and are routinely exposed to insecticides and pesticides. Beatings by farm owners are commonly reported.

The environment also suffers from unsustainable cocoa cultivation. Rain forests have been cleared to satisfy our demand for chocolate – 14 per cent of the Ivory Coast's rain forests had been clear-cut by 2000 to allow cocoa farming. This is not the traditional method, in which cacao trees grow in the shade of larger trees, but modern 'sun cultivation', which requires the use of larger amounts of pesticides. Frequently, commercial cocoa growers fail to rotate their crops, further reducing the region's biodiversity. Rain forest deforestation and single-crop farming threaten many endangered species. Meanwhile, pesticides and other chemicals poison the air and water as well as encouraging the development of pests and diseases resistant to pesticides.

The Fair Trade Association was established to address these problems, to ensure that farmers are fairly paid and to counter the harm done by unsustainable cocoa cultivation. The Fair Trade Association stipulates that farmers will be paid a fair minimum price for cocoa beans or, if the world price rises above this established level, that the price paid to farmers will be adjusted upward accordingly. The association also demands that no child labour or forced labour is used in cocoa production.

Although the Fair Trade Association does not require farmers' products to be certified as being grown according to organic standards, it does require "integrated crop management" and bans several pesticides. Farmers are guaranteed a few pence more for certified organic cocoa beans, thus giving them an economic incentive to farm organically. Further, Fair Trade Association cooperatives are required to reserve some of their profits to finance workshops where farmers learn the ecological value of organic and sustainable farming systems.

Coffee, tea, and thee

If you like coffee with your chocolate, drink organic, fair-trade coffee. It's the ecological choice, and not only ensures worker safety but also safeguards a vital habitat for migrating birds. Modern coffee-growing practices decimate bird species. Traditionally, coffee was grown under a canopy of trees that provided a home for diverse species, but newer coffee-growing techniques rely on sun cultivation, which produces biological deserts. Sun cultivation produces a higher yield and more profits, but also requires more chemicals and harms biodiversity. You can have a direct influence on growing methods by asking for organic, shade-grown beans – which coffee connoisseurs consider far superior in flavour to the industrial product, anyway.

If tea is your favourite beverage, drink organic and herbal varieties. Remember, too, that loose tea means less packaging. Make a pot and share it with a friend. You can easily grow herbs for tea – camomile, lemon verbena, mints – and brew with fresh leaves. Dry them for use during the winter. I keep a separate pot for fragrant herbal teas.

Ordinary black tea, which is harvested from the tea plant – a shrub in the camellia family that originated in India – contains a range of tannins, which function as antioxidants. Researchers are analysing these tannins to determine their potential health benefits. Green tea, harvested from the same plant as black tea but dried rather than fermented, helps protect against cancer and other ills, while specially prepared white tea contains three times the amount of antioxidants found in green and black teas, because it is dried in sunlight rather than in an enclosed space.

Bottled water: H2no!

Most of the 700 chemical contaminants in public drinking water are not detectable to the nose or eye, but the standard of public water suppliers in developed countries is high. Bottled water may taste good, but it's bad news for the environment because of the energy used to process and transport it. Tests show that bottled water is no purer than water from your tap – and it's sometimes

less pure. If you must buy bottled water, choose large glass bottles – water bottled in plastic absorbs a certain amount of carcinogenic polymers – and recycle them. If you like to carry water, re-use a bottle or buy a good-quality water bottle at an outdoor sports shop and refill it often. Keep filled bottles in the fridge for a quick exit. A stainless steel vacuum flask keeps water icy cold on even the hottest day.

Public drinking water does present some problems, many of which are related to high-chemical agriculture. Potential cancer-causing disinfectant by-products (DBPs) are found in some water supplies. DBPs include trihalomethanes, formed when chlorine reacts with organic matter, such as decaying leaves.

For the sake of safety, use cold tap water for cooking and drinking, and flush the pipes with cold water for several minutes before using water in the morning. If your water smells of chlorine, you can leave some in an uncovered container in the fridge overnight to let most of the chlorine vaporize.

Many water authorities and scientists discourage the use of home water filters because bacteria can multiply in them if they are not well maintained. However, in some areas tap water tastes distinctly unpleasant and filtering is the only way to make it palatable. Filtering does not remove all the impurities, but it can provide safer, more pleasant tasting water at a reasonable price. If you use a plastic jug filtering system, change the disposable filter often; otherwise, it can release heavy metals and bacteria. A plumbed-in filtering system is cheaper in the long run and has fewer disposable parts.

THE GREEN CHEF — CONSERVING FLAVOUR AND ENERGY

When you cook, try to recapture the joy that children experience – the sense of fun and experimentation. Too many of us feel overwhelmed in the kitchen, but that's often because we think we should be able to produce the same quality that TV chefs do. The single biggest aid to culinary success is having the right guide, which means the right cookbooks – and there are many to choose from – and the right companion with whom to share the adventure. The right equipment is important, but you don't need much to create the delicious foods that bring your friends and family together.

Here is a list of basic equipment:

- Two good-quality, well-sharpened knives – one small, one large
- One large frying pan that distributes heat evenly
- One covered saucepan
- One really big pot to give pasta or soup the space it needs
- A large sieve or colander
- At least two chopping boards – a large one for most foods and another solely for meats (a small piece of scrap pine is perfect) – and if you want to be fancy, another for bread
- Wooden spoons for stirring, a spatula, a ladle and a slotted spoon

And here are some tips for reducing your energy usage:

- A microwave oven uses only a fraction of the energy other cooking appliances use, especially if you're heating a single dish. Reheating homemade dishes rather than preparing prepackaged foods is also a good eco-choice.

- Cooking in the oven uses more energy than any other method. A toaster oven is a good alternative, and some casseroles and even baked potatoes can be prepared in it. Serious cooks tend to prefer gas to electricity, and gas is cheaper and thus more energy efficient.
- To reduce cooking time drastically, use a pressure cooker. Make it stainless steel rather than aluminium for safety's sake.
- Chop vegetables, including potatoes, into small pieces – they'll cook much more quickly.
- Let dishes cook in their own heat – bring spaghetti sauce or soup to the boil, then turn off the flame and allow the pan to stand rather than simmer for a couple of hours.
- Use a tiered steamer that will cook several types of vegetables at once.
- Use an automatic switch-off electric kettle to heat water, and heat only as much as you need.
- Scale deposits will make your kettle less efficient, so from time to time clean it with a strong vinegar solution.
- Pour unused boiling water from the kettle into a vacuum flask for later or use it to start soaking a pan of beans.
- Improve your health and save energy by eating more raw foods, fresh fruits and salads.

- To reduce cooking time, cover cooking pots.
- Size the fire to fit the pan. A flame that is licking up the side of a small saucepan is wasted, and also makes a mess on the outside of the pot.
- Don't turn on the whole grill to make a single piece of toast – use a toaster.
- Fill the oven – roast some vegetables in a pan on the lower shelf while bread or biscuits are baking on the top shelf.

The green chef's appliances

Chest freezers are more efficient than upright models, although less convenient to use. If you have a freezer, defrost it regularly and try to keep it full. Keeping it outside the house, in a garage or cellar, reduces energy costs. Otherwise, put aluminium foil on the wall behind it to reflect waste heat into the room.

Dishwashers save time for a large family and certainly keep the kitchen tidier, but they require strong detergents – read the warning label – and use large amounts of hot water as well as large amounts of energy in the dry cycle. On the other hand, if operated on an economy setting with full loads, a dishwasher often uses less water than hand washing. If you use a dishwasher, use a green phosphate- and chlorine-free detergent and reduce the amount you use to a minimum. Choose the economy setting and turn off the dishwasher when it gets to the dry cycle – open the door and pull out the racks to let the dishes air dry.

Refrigerators are more efficient if you keep their seals clean and vacuum the metal fins on the back regularly. If you have a fridge that is more than five years old, consider a new energy-saving, frost-free model. Ensure that your old fridge is recycled – fridges contain ozone-damaging chlorofluorocarbons (CFCs).

A potluck for the planet

We need to work together to solve the social and environmental problems, and what better way to build connections than to eat together? All you have to do is master a single recipe. Make a big batch of whatever it is and then invite your friends and colleagues to come over with their favourite dish. Enjoy delicious food while coming up with ideas for shared green projects, from a bicycle club to an annual street clean-up. Call it a potluck for the planet!

CHAPTER FOUR

HEALTHY YOU, HEALTHIER WORLD

One of the mysteries of modern life is how we can have it so good and yet feel so bad. Health problems that plagued our ancestors – from poor teeth to diseases that wiped out whole villages and infections that killed young mothers within days of giving birth – don't worry us at all. But how many people do you know who are bursting with good health? Too many of us suffer from the diseases of civilization, such as back pain, obesity, and stress-related headaches.

Many experts believe that the increase in certain cancers, especially among younger people, is due to exposure to chemical substances that are now loose in the environment. We also have a host of new health worries, from the recurrence of tuberculosis (TB) to multiple chemical sensitivity. We're not so different from battery-reared chickens, living under artificial lights, breathing stuffy air and not getting enough exercise.

SELF-CARE BASICS

Redefine yourself as your best health-care provider! You can do more than any doctor, allopathic or osteopathic, to deal with your stress and fatigue and lack of fitness and to recognize when something might be seriously wrong with you.

Here are some tips for basic self-care:

- Be proactive with health-care professionals – ask questions and get involved in your diagnosis and treatment options.
- Drink alcohol only in moderation, and don't smoke.
- Maintain your weight at a healthy level.
- Get nutrients from the foods you eat, not from supplements.
- Don't try to medicate the effects of stress. Take a short walk or nap, or drink a cup of peppermint, camomile, or sage tea.
- Get lots of exercise.
- Get plenty of fresh air, sleep and natural light.
- Surround yourself with people and things that you love.
- At home and at work, reduce your exposure to pollutants.

Avoid allergies

Nowadays allergies are increasing, and many people believe that they are allergic to some household substance or food. That's not surprising when you look at the astounding increase in the number of new chemicals that has come into use during recent decades. However, allergic reactions are also connected to our individual immune responses, and more of us are sensitive to natural substances, such as pollen. Asthma and hay fever also appear to be on the increase. In fact, cases of childhood asthma have reached epidemic proportions in the last 15 years or so.

Allergies are an overreaction of our immune systems – our bodies' defence mechanisms – to irritating stimuli that our bodies usually handle without trouble. Researchers suggest three causes for the increase in overreaction.

1. We are living in far more hygienic surroundings than at any time in the past, and as a consequence, babies fail to develop sufficiently robust immune systems.

2. Babies are not adequately breast-fed. Despite the UK government recommendation that babies be fed with breast milk only for the first six months of life, more than 21% of British mothers give up within the first two weeks and 36% within the first six weeks. The proportion of mothers breast-feeding at two weeks is 52%, dropping to just 13% at nine months. Breast-feeding is an important part of the process of developing a strong immune system because it allows a baby to build his or her own immune system with help from mother's antibodies. Breast-fed babies are substantially healthier as infants and generally healthier as adults. There is a reduction in later childhood obesity by 15-20 per cent for infants that have been breast-fed six months.

3. The pollution in our environment and the number of human-made chemical products to which we are exposed create an additional source of stress for weakened immune systems.

Reduce the chemical burden

Pollution and environmental change have always harmed human health. Historically, diseases spread when people moved into early cities that had inadequate waste disposal systems, and our ancestors also became sick because they breathed too much smoke from wood and coal fires. Modern life deals us a different hand. Today, instead of smoke and contagion, we are exposed to a wide range of human-made chemicals – and they may cause headaches, allergic reactions and long-term health damage. Every day new chemicals are being developed and put on store shelves without adequate safety testing.

Staying indoors won't prevent us from being exposed to such chemicals. Ironically, the air inside our houses and offices is usually far more polluted than the air outside, because of paints and varnishes, office machines, modern building materials, soft plastics, cleaning products and so-called air fresheners.

Did you know? Our bodies contain traces of 500 chemicals that did not even exist in 1920.

Did you know? Chlorine bleaches and waste incineration are major sources of dioxins in the environment.

Green sex — don't knock it till you've tried it!

More than six billion people inhabit the planet now, and overpopulation will continue to be a major problem in the 21st century. Thus, contraception is an essential part of eco-living, in spite of concerns about the health effects of both spermicides and the pill, and environmental issues such as disposal of condoms.

We've been trying to prevent unwanted pregnancies for thousands of years. Many people advocate natural family planning (NFP) or 'rhythm method' for both religious and health reasons. However, NFP often produces a large natural family! On balance, it's better to choose a form of contraception that makes sense on the basis of your lifestyle, not on it's being "natural."

No contraceptive is foolproof, but new low-dose birth control pills are very effective and seemingly protect against certain forms of cancer, too. Conventional reusable diaphragms are a good environmental choice, but many couples find the necessary spermicides unpleasant to taste and smell and also a health worry.

Condoms have the advantages of both protecting against sexually transmitted diseases and being widely available. However, while they were originally made from biodegradable animal tissues, they are now made from almost indestructible latex, and their packaging rivals that of compact discs for eco-unfriendliness. Flushed condoms will pollute beaches and harm wildlife – birds think that condoms are edible – so be sure to throw them in the bin, not flush them down the toilet.

INTEGRATIVE HEALTH

Alternative medicine is now widely accepted – and some GPs now routinely refer patients to osteopaths, acupuncturists and hypnotherapists. With public demand for their services increasing, complementary practitioners are becoming a normal part of modern health systems. The GP is the obvious point of referral for patients who wish to access an appropriately trained and motivated practitioner. These are listed in a dedicated NHS Directory (http://www.nhsdirectory.org) to provide access to all practitioners, who have put themselves forward to work either directly in NHS practices or from their own practice on a referral basis.

Take fewer antibiotics and other drugs

We can clearly see the human disease ecosystem – our connections with other people – in the growing global problem of drug resistance. This occurs because antibiotics are either over-prescribed or improperly taken by patients. Legal prescription drugs constitute one of the most profitable businesses on Earth, and professional praise – to say nothing of huge grants – are heaped upon researchers who develop expensive, resource-intensive medical 'miracles'. Naturally, these miracles usually result in increased profits for the pharmaceutical companies.

Lighten up

We are becoming much more aware of the influence of light on our health. During the winter months, some people suffer from a condition called 'seasonal affective disorder', or SAD. This is thought to be caused by overproduction of the hormone melatonin. The symptoms of SAD are severe depression that lifts when spring arrives, weight gain, excessive sleeping, and cravings for carbohydrates. Therapy includes sitting in front of huge sets of full-spectrum lights. A more natural therapy – one that all of us can use for these winter blues, is to spend more time outside. Spend a minimum of 15 minutes a day outside in summer and 30 minutes in winter, when the sun is less intense.

TIPS FOR TAPPING THE SUN

* *Don't wear sunglasses all the time. Your eyes need some sunlight unfiltered by window panes or lenses. But do remember to wear sunblock!*

* Work close to a window and open it when the weather is fine.

* *Cycle or walk instead of driving.*

* Participate in sports and other outdoor activities in preference to indoor activities.

GET UP, GET OUT, GET MOVING

Environmentalists haven't talked about sports until recently. And some seemed to equate a passion for tennis or football with Nero's fiddling while Rome burned. Nor have environmentalists talked much about how we spend our leisure time, except occasionally to criticize competitive games at school.

However, sports are a major business globally and have a significant impact on the environment. Exercise also is one of the most important facets of an environmentally sound lifestyle because it involves you in the world and makes you aware of your body and your surroundings. Moreover, regular exercise is probably more important than diet in keeping you healthy, and is essential to building a strong immune system that can cope with all the challenges that today's world throws at us. Exercise gives a new meaning to the term *eco-activist*!

Did you know? A golf course in Thailand uses as much water every day as 60,000 villagers.

A simple regimen of regular exercise has significant physical and psychological benefits. Exercise boosts energy levels, tones muscles, decreases arthritic pain and chronic backache, and has antidepressant effects. Even a moderate amount of walking dramatically improves women's chances of recovery from breast cancer. Experts recommend walking for people who suffer from

psychological disorders, including chronic fatigue syndrome or myalgic encephalomyelitis (ME) and severe depression. So just imagine what exercise can do for those of us who suffer from job stress and ordinary inertia!

Throughout history, the human race has been trying to find ways to reduce physical labour. We're still doing it – with TV remote controls that let us avoid having to get up to switch channels and electric lights we can turn on with a clap of our hands from across the room. It's all eminently convenient. But the trade-off is that throughout the developed world, people are out of shape. It's no mere coincidence that the United States – the land of the free and the home of the self-propelled mower – is the fattest nation on Earth and is packing on more pounds every day.

Biologically, people are meant to be active. We're built that way. Although most of us cannot change the sedentary nature of our jobs, we *can* re-orient our leisure time to be more active. We'll be happier and healthier and far better able to cope with unavoidable exposure to pollutants in water, air and foods.

The eco-aware person looks for ways to enjoy and explore the natural world by being active outside – *in* the natural world. Physical activity, at its best, reconnects us with our physical being and, by extension, helps us understand ourselves as part of the natural world. It's exhilarating to climb to the top of a hill and survey the landscape below or to feel your calf muscles burn after walking or running that extra mile.

TIPS FOR SELECTING
SPORTING ACTIVITIES

Choose non-power-assisted sports – a canoe or a bicycle, say, instead of a motorboat or a jet ski – because they are both easier on the planet and also more sociable and stress-relieving.

Choose cross-country skiing instead of downhill skiing, which has considerable ecological impact – both because of snow-making machines and because it requires long air or car trips to get to the slopes.

Think twice about installing a swimming pool at home, even if you can afford it. Home pools have a significant negative environmental impact because of their water and energy consumption.

Look for leisure activities that you can enjoy close to home – without driving or flying long distances.

Did you know? Support networks – our community – really do affect health. People who join just one new social group or club are 50 per cent less likely to die in the next year, according to leading social capital expert Robert Putnam.

Not so long ago most people walked a lot – to do the shopping, catch a bus, or visit a relative. However, as cars have proliferated, we walk less and less. So do our children. But with a little reorganizing, we could start walking again and reap many health benefits while also reducing wear and tear on planet Earth.

To get the most benefit, walk steadily for at least 20 minutes three times a week. Walking can be done year around. Your equipment needs are few: just a comfortable pair of shoes and appropriate clothing. I always wear sunblock on my face and neck.

Cycling is another enjoyable way to exercise, although cycling isn't always as efficient as walking because it's not as steady – freewheeling down hills doesn't require any effort. As cycles are so efficient at converting effort into motion, it's also difficult to go fast enough to elevate your heart rate to a beneficial level.

Running, in terms of low-tech aerobic activities that fit into almost anyone's lifestyle, gets a high score. However, it's not a practical way to get around because you can't carry things easily and you are sweaty at the end. Also, injury rates from running are much higher than from walking.

Obviously, exercise is much more pleasant if you can breathe clean air as you exercise. In a city, try to exercise early in the morning and on weekends, and, of course, stay off the main roads. Some cyclists wear masks to prevent inhaling particulates from car exhausts. Whenever possible, exercise away from industry and cars and remember that the air inside a gym may be just as polluted as the air outside, although the benefits of exercise in a gym far outweigh exposure to a little more polluted air.

BODY CARE — MORE THAN SKIN DEEP

Go into your bathroom and look at the labels of the products. How many labels list 'aqua' as the first ingredient? That means you've paid all that money for plain water. Because transporting goods is a source of much pollution – and contributes to global warming – moving water around the globe is a serious problem. The best favour that manufacturers could do for the environment would be to sell concentrated products for us to mix with water in our homes. But manufacturers worry that we wouldn't pay all that money for a bottle of skin toner if all we could see was the meagre teaspoonful of ingredients left after the water was extracted. Also, some products obviously need to be mixed and emulsified before they are bottled.

Ideally we'd be able to refill our own plastic or glass bottles in a local shop. Some health shops do sell cleansers and shampoo in bulk so that customers can bring their own containers.

Did you know? "Wild-harvested" is a popular adjective in body-care products and some foods, and it sounds good, suggesting pure nature. But did you know that the wild harvesting of plants – as well as of ocean fish, of course – threatens them with extinction?

TIPS FOR BODY CARE

- *Mix your own simple beauty preparations – see books or the Internet for recipes.*

- Carry a handkerchief instead of tissues – it's easier on the skin, too.

- *Do not drop cotton buds or anything except toilet paper and flushable tampons into the toilet.*

- Purchase concentrated and lightweight products.

- *Choose products that come in large sizes and simple packaging.*

- Purchase refillable products.

Skin-care products

If you like natural, organic skin-care products and can afford them, go right ahead. But if your budget is limited, don't feel guilty about using basic inexpensive products, natural or not. Skin-care products have relatively little environmental impact, and it's better to direct your eco-conscience into other areas.

Most ingredients in sunblock are not natural, but they are important because the thinning of the ozone layer allows more dangerous ultraviolet (UV) radiation to reach us. The rate of the fatal skin cancer, malignant melanoma, is especially high among people of northern European extraction, but even people with darker skins need sun protection.

In addition, you should wear a sunblock that contains zinc oxide, titanium dioxide, or avobenzone as an active ingredient to protect you against UV rays. In advertisements for tanning salons, the longer UV rays have been advertised as safe tanning rays because they develop melanin particles in the skin and tan the skin without burning it. But recent research shows that, in the long term, UV rays are far more damaging because they cause skin cancer and premature and irreversible aging of the skin.

TIPS TO PROTECT YOUR SKIN

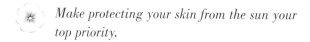

Make protecting your skin from the sun your top priority.

Apply an expensive skin cream judiciously to make it last for many months.

When possible, use vegetable-based – rather than petroleum-based – products. A jar of vegetable-based jelly, made for babies and sold in whole-food shops, is almost identical to expensive lip salves.

Wear a moisturizer containing sunblock *every* day, all year around. As much as 80 per cent of the sun's radiation penetrates the atmosphere and reaches Earth. Be sure that your sunblock's sun protection factor (SPF) is at least 15 *and* that it contains zinc oxide, titanium dioxide, or avobenzone. A sunblock with only a high SPF is not enough protection.

Shop around if your skin reacts badly to the common sun-block ingredient para-aminobenzoic acid (PABA). It's possible to find excellent products that don't irritate sensitive skin.

If you are swimming or sweating, re-apply all creams frequently.

Supplement your sunblock with a beach umbrella, a hat and a silk wrap over your shoulders. But remember that sunlight can penetrate penetrating sheer clothing, and also that it's reflected from water and pavements.

Most health experts recommend that we avoid sunbathing for hours on end in pursuit of a tan. If a moderate but traditional tan is what you want, you could try a self-tanning lotion, coupled with a little genuine sunshine. However, no one knows whether these products have long-term health risks. So, as with any chemicals, be cautious.

To avoid eye problems later in life, consider wearing sunglasses if you are in the sun a lot. Be sure that the lenses absorb 100 per cent of both UVA and UVB light.

Examine your skin for growths, itchy patches, sores that won't heal, and any changes in moles or coloured areas. Such irregularities could be cancerous. If detected early, skin cancer can almost always be treated successfully.

Cosmetics

Most cosmetics are outrageously over-packaged, and many cost a fortune. The amount of actual cosmetic in the product is small and the price is large, but the packaging makes the exchange seem less unreasonable. A few cosmetics companies sell refills, which are a good idea, but on the whole, consumers have little choice except to buy the largest size possible and complain to the cosmetics companies.

More and more cosmetics companies are becoming aware of the skin sensitivities of customers and are eliminating irritating ingredients, particularly fragrances. Some standard ingredients in lipsticks, such as plastics, mineral oil, saccharin, and artificial colours, have been shown to cause cancer in animals. Choose unscented cosmetics that have not been tested on animals and avoid artificial colours and fragrances. Also avoid cosmetics that contain possible hormone-disrupting chemicals such as octoxynol, nonoxynol, and nonylphenol ethoxylate.

Nail care

Nothing is wrong with wanting your hands to look pretty, but if your fingernails get in the way of gardening or playing tennis, think again! Try alternatives to nail polish, which contains formaldehyde resin, and remember that acetone solvents can cause skin rashes, dry out your nails and, as an unexpected bonus, dissolve plastics! If you do apply nail polish, be sure to do it in a well ventilated room. As an alternative, try buffing your nails. Buffing gives a good-looking shine without exposing your fingers to chemicals.

Eye care

Eyeglasses are more eco-friendly than contact lenses, because they require no tablets, solutions, or sterilization. But those benefits are not going to persuade many people to abandon contacts for glasses. If you wear contacts, ask for lenses and a lens-care system that are as simple as possible. Manufacturers make tons of money on the daily solution tubes and tiny bottles. The cheapest safe lens-care system will be easiest on the environment, and a good professional will be honest about what you really need.

Deodorants and antiperspirants

Sweating is a healthy and natural bodily function, and women's growing interest in being physically active is beginning to change the dainty, no-pores ideal of our mothers and grandmothers. But try not to use antiperspirants unless you are wearing very delicate or fitted clothing. For ecological body-odour control, use roll-on or pump-action deodorants. New commercial products use natural bactericides and odour reducers such as lichens, coriander, lemon and tea tree oil. Minerals are also extremely effective. You can use a deodorant crystal or baking soda – keep some in a pretty jar and pat it on with dampened fingertips.

The kindest cut of all

If you want to get an extremely clean shave, use a standard razor that has replaceable blades rather than an electric razor or plastic disposable. Ecologically speaking, an old-fashioned straight razor would be even better!

Hair care

If your hair doesn't shine with good health, look first at your general health and stress level. Your diet also may be deficient in the right fats – eating more raw nuts and switching to cold-pressed soy and virgin olive oil could make your hair more shiny. Regular scalp massage and thorough brushing can also help. An amazing number of shampoos are medicated – just look at the shelves. But here again, alternatives are a wise choice. For example, anti-dandruff shampoos contain toxic chemicals, primarily selenium sulphide, which can, if swallowed, cause the degeneration of internal organs. Resorcinol, which is easily absorbed through the skin, is another dangerous chemical that's a common ingredient in shampoos. A seaweed-based shampoo has proved to be a successful dandruff treatment, and tea tree oil shampoo is amazingly effective against dandruff, too.

Hairspray

Hairspray often contains toxic resins and propellant gases. Finely dispersed in an aerosol, these resins and gases can cause a lung disease called thesaurosis. Instead, use a pump spray or just rub gel into your hair.

Drying your hair

Electric hair-dryers emit strong electromagnetic fields because they use a lot of energy. They can also damage hair. Your hair, your body and your planet will be healthier if you have a hair style that is quick to dry or can dry naturally.

If you do use an electric hair-dryer, don't use it when your hair is still soaking wet. Towel dry and then air dry your hair for a little while; use the dryer only for the final styling.

Dryers work more efficiently if you clean the intake vents every month with an old toothbrush.

Colouring your hair

Chemicals in commercial hair dyes are known to cause birth defects and are suspected of causing cancer. Because your scalp can absorb these chemicals, consumer advocates caution pregnant women not to use them. Natural colour is becoming more fashionable, but if you do colour your hair, choose a safe plant-based product. Ask your hairdresser to recommend one. You can also try homemade mixtures. They range from lemon juice and camomile tea to lighten fair hair to black coffee or walnut shell extract to cover the gray in dark hair. If you must use chemical dyes, use them exclusively as highlights to your natural colour so you can avoid having the chemicals touch your scalp.

Sanitary protection

- Look for organic cotton tampons in whole-food shops. Unfortunately, these tampons are expensive – but prices are coming down.
- Consider using non-disposable sanitary protection. The modern products – sponges, cups and cloth pads – are a far cry from the simple rags that our great-grandmothers used.
- Use tampons that are made from unbleached cotton and biodegradable cardboard.
- Avoid using tampons that contain ultra-absorbent materials because they promote the growth of the bacteria that cause toxic shock syndrome.
- Never flush sanitary pads down the toilet.

YOUR CHILDREN'S HEALTH

Pregnant women and their unborn children are particularly vulnerable to chemical toxins. A baby is affected by everything a pregnant woman consumes, breathes, or touches. Members of hospital operating room staffs, for example, suffered a very high miscarriage and infertility rate because of their exposure to anaesthetics. Fortunately, this has led to improved ventilation systems in operating rooms.

Because children take in more air and more food in relation to their body weight than do adults, they are more sensitive to environmental toxins. Children's reactions to chemicals can include hyperactivity and even psychiatric disorders. In the United States, regulations are being initiated to provide special protection to children.

Did you know? Solvents, PCBs, lead and heavy metals such as cadmium and mercury damage the foetal brain.

TIPS FOR PREGNANCY – AND CONCEPTION

Limit your intake of coffee and tea.

Avoid alcohol or lower your intake dramatically.

Avoid all medications, even over-the-counter drugs, without specific guidance from your physician.

Avoid food additives, including artificial sweeteners.

Give up hair colouring during your pregnancy, or switch to vegetable-based treatments.

* Eat organic foods and lean meat, because hormones, pesticides and antibiotics concentrate in fatty tissue.

* *Finally, don't paint the nursery at the last moment. When the baby moves in, the room should smell of fresh air and sunshine, not paint fumes.*

Baby care and babywear

The best baby clothes, from an ecological standpoint, have been worn before. Jumble sales offer terrific bargains, and if you're lucky, relatives and friends can give you plenty of gently worn clothing to kit out your child for the first year or two.

Not so long ago, hospitals handed out free samples of disposable nappies, giving them tacit medical endorsement. But with landfill space at a premium today, some hospitals are advertising services that deliver fresh cloth nappies and pick up the dirty ones about twice a week. Although the initial expense of cloth nappies is greater than the expense of a bag of disposables, in the long run, cloth nappies cost only half as much, even when you take laundering into account. But even more to the point, babies with sensitive skin tend to get far fewer rashes if they are wearing cloth nappies.

Here are some tips for green child care:
- Do not buy toys made from soft plastics because these materials contain toxins such as phthalates, salts or esters of phthalic acid that are used as plasticizers and in solvents.

Greenpeace has led an international campaign against these unsafe playthings, especially teething toys that are designed for children under three.

- Buy for durability. Toys made of solid natural materials are expensive, but they can be passed down to friends or grandchildren or sold through small ads.
- Buy toys and books at jumble sales or in charity shops.
- Check your local free sheet as well as the notices on community bulletin boards and in shop windows for second-hand baby clothes or place your own advertisements.
- Sign up with a nappy service or buy three to four dozen cloth nappies and covers.
- Stock up on nontoxic cleaning products.
- Wild places are amazingly significant in in the development of children's personalities. Take family holidays where your children can climb trees and build forts. You can also plant hideaways or build forts, dens or tree-houses in the garden for your children to explore and enjoy.

Active children

Active children are better tempered, more alert and healthier. As adults, they will be less likely to develop osteoporosis, heart disease and other ailments. Experts recommend at least an hour a day of vigorous activity for children. Cycling or walking to school helps your child develop a sense of independence and physical confidence. You should get your pre-school children walking, too. Take your child out of the baby buggy and walk hand in hand so that the two of you can look into each other's face, talk, examine leaves and have a chat with a squirrel! After-school physical activities are also important. Tree climbing, active games and garden chores develop children's strength and motor skills.

HEALTHY PETS

Reducing the amount of meat that we eat is one of the most important steps that we can take to reduce air pollution and global warming, so it's unfortunate that the pets we love most are generally carnivorous. Other animals are raised and slaughtered just to feed them. Manufacturers spend millions every year to advertise pet foods, and the range of products designed to tempt our fussy companions is growing.

Depending on the animal, approximately ten times as much vegetable or grain protein is needed to produce animal protein, and the grain that is used to feed the livestock is often imported from Third World countries. In addition, 80 per cent of pet owners buy canned pet food for their companion animals. The packaging and energy that is used in processing, packaging and shipping branded pet food and other pet products adds considerably to their environmental costs.

Many commercial pet foods are the equivalent of junk food – addictive and appealing but an environmental and nutritional disaster, full of sugar, fat, salt and fillers. Whole-food stores and vets sell higher-quality, additive-free pet foods. The best-quality dried pet foods are a complete diet and are thoughtfully packaged in cardboard or paper. Dry pet foods are also lighter, which lowers both transportation costs and energy use.

TIPS FOR GREENER PET-KEEPING

* *Ask the whole-food store or your vet about nontoxic alternatives to flea collars. Most of the flea collars on the market contain potent insecticides, so it's important to keep them away from children.*

* Prevent your pet from chasing or killing birds and wildlife.

* *Consider a vegetarian diet for dogs – though this is not a safe alternative for cats – and use dried food because it has less packaging and weighs less.*

* Think small. Big dogs are comparable to big cars in terms of environmental impact and expense. Small dogs also cope better with urban environments.

* *Limit yourself to one or two pets.*

* Don't allow your pet to reproduce, however much you may love kittens and puppies.

CHAPTER FIVE

GREENER WORKPLACE, CLEANER PLANET

The place where you work is what sociologists call your 'second place'. Home is, of course, the first place in our lives, while the 'third place' is where we go to socialize – in coffee shops, bars, exercise clubs and even barber and beauty shops.

Many of us spend a majority of our waking time in the workplace, the second place, and that's one reason why it merits special attention. Also, increasing millions of people working at home in westernized countries are changing their homes into offices as they devote all of their time and resources to developing their businesses.

You can make your home greener and healthier, but unless you also make your workplace greener and healthier, you may still suffer from ozone pollution from photocopiers and printers and a stiff neck from a poorly positioned computer screen.

Today's progressive business leaders know that taking corporate social responsibility includes reducing waste and creating a livable environment for their employees. They also know that instituting green policies will save them money in the long run and make their employees more productive and more engaged.

Businesses consume a huge amount of natural resources and create enormous waste. For instance, the average office worker uses four to nine kilograms of paper each month. Just imagine how much good it would do if the employees of an entire department or an entire company followed the lead of one green co-worker who made a few sensible changes.

THE DIGITAL DREAM

When the computer age dawned, many of us embraced it because we believed it would decrease the amount of paper we used as well as the number and frequency of business trips. But reality has not matched that vision. The more computers we have, the more paper we generate; the more businesses we have connected by LANs (local area networks), WANs (wide area networks), and Intranet sites, the more trips we take. Frequent flyer mileage points have become a kind of global currency.

On the other hand, our ability to communicate globally and to share information via the Internet can be a great tool for creating a healthy environment for everyone on Earth. That ability enables fast, cheap communication as well as sophisticated monitoring of current problems and modelling of future environmental scenarios.

Modern communications enable more and more people to work in their homes – for example, operating their own graphic design business or telecommuting. More and more professionals spend one or two days each week working from home. As a result, companies benefit by having more productive employees. Telecommuters, their families and communities, and the environment also benefit.

These new technologies also allow people to relocate their businesses to rural areas, so people can be employed without contributing to the traffic or noise in cities. Considering the pressure on housing in already densely populated areas, this kind of relocation can be a true benefit, but human factors continue to weigh in. Businesses still want their employees in one place, and people naturally cluster where job opportunities are plentiful.

Imagine... Innovative companies, including giants like Xerox, have begun to supply services rather than goods – if you're in a position to choose, you can lease equipment and even carpets. This gives companies an incentive to design things to last and to use parts that can be re-manufactured.

THE DIGITAL DILEMMA

During its early days of development, most people thought that the computer industry was clean, that it did not flush torrents of sludge into rivers or belch clouds of smoke into the sky. But sadly, the industry has a dark side: toxic waste. California's Silicon Valley, home of the computer industry, has the highest concentration of toxic waste sites in the United States.

Several hazardous materials are used in the world-wide manufacture of computers, putting people who work at production sites and people who live nearby at risk. Typically, these risks include increased rates of cancer, birth defects and neurological disorders. The average life of a computer is about two years, and few are reused. When thrown out and dumped in land-fill sites, they create hazardous electronic waste (e-waste). As an example, 250 million obsolete computers generate a huge amount of e-waste:

- 181 tonnes of mercury
- 544 tonnes of chromium
- 861 tonnes of cadmium
- 450,000 tonnes of lead
- 1.8 million tonnes of plastic

Some large international corporations have Computer TakeBack schemes and accept back used computers so they can be de-toxified. But discarded personal computers tend to end up in land-fill sites, and most of the e-waste of developed nations is shipped to Asian nations for processing.

Poor Asian communities consider e-waste a resource, but bear the environmental consequences of processing it, because toxins are allowed to leach into the ground and water supplies.

The statistics of our disposable digital world are alarming:

- More than 375 million used printer cartridges are thrown into landfills each year.
- One laser printer cartridge contributes approximately a kilogram of plastic and ink to landfills.
- Manufacturing one laser printer cartridge requires almost 3.5 litres of oil.
- In the US, 130 million mobile phones are discarded each year.
- Computers and cell phones contain potentially toxic elements such as zinc, nickel, lead, copper, cadmium, beryllium, arsenic, and antimony.

Message it! Let computer and software manufacturers know that you – and your company – think the environmental effects of computers should be the focus of their corporate social responsibility programme. They shouldn't boast about their support for AIDS research while they're ignoring the pollution from their own industry!

The heavy metal lead is an especially worrisome component of computers and mobile phones. The typical cathode-ray tube computer monitor contains two to three kilograms of lead.

According to the CTBC, lead can cause brain damage and developmental and reproductive problems and is also a suspected carcinogen. Cadmium is a known carcinogen and can also cause liver and kidney damage.

Microchips are the brains of computers and cell phones. But these tiny items are big consumers of resources. To manufacture a single silicon wafer containing 240 microchips requires:

- Almost 300 kilowatt-hours of electricity
- Chemicals weighing nine kilograms
- More than 8,300 litre of de-ionized water
- Hazardous gases totalling well over 500 litres
- More than 10,000 litres of waste water

These statistics are a part of the problem. How can you be part of the solution? First, support the concept of extended producer responsibility (EPR), an extension of the 'polluter pays' principle. Computer manufacturers signing up to EPR take full financial and physical responsibility for their products from manufacture through disposal, refurbishment, or recycling. Fewer than ten per cent of outdated computer products are recycled or refurbished. Until an effective recycling programme exists, companies will have no incentive to build recyclable computers.

Second, look for an energy efficiency label on the next computer you buy. Energy-efficient computers in their sleep mode use 70 per cent less electricity than computers without this power-saving option.

Third, look for eco-labels among computer brands. The Blue Angel label in Germany, the Swan label in Scandinavia, and the TCO label in Sweden are examples. NEC's PowerMate ECO desktop computer contains no lead, polyvinyl chloride (PVC), mercury, cadmium, chromium, or boron. It has a flat-panel monitor screen, is made from recyclable plastic, and is Energy Star compliant. Fujitsu has heralded its biodegradable personal computer casings made from vegetable resins.

Fourth, ask if the computer contains polybromated diphenyl ethers (PBDEs). PBDEs are suspected to be endocrine system disruptors and research has shown that they can interfere with brain development and may cause cancer in laboratory animals.

Did you know? The average person buys a new mobile phone every 18 to 24 months.

The high-tech horsepower race – our incessant quest for more RAM, bigger hard drives, faster modems – helps create the problems associated with the manufacturing and disposing of so many computers. Every time the processing speed of computers doubles, millions of consumers rush out to buy new cutting-edge machines and junk their old ones.

Relatively speaking, only a few people are able to afford the newest and biggest and fastest. This disparity contributes to the growing gulf between the have and the have-not nations.

CREATE A GREEN OFFICE

Research has shown that as many as 70 per cent of employees suffer health complaints caused by their work environment. Some of these complaints are caused by working in modern buildings – sick building syndrome – and others are caused by working at poorly arranged workstations and by spending too much time at the computer.

Here are some steps that employers and employees can take to create a green office:

- Encourage employees to cycle to work by providing bike racks and showers.
- At lunchtime, instead of going out with colleagues for a leisurely drink, go out for a brisk walk.
- Offer, or ask for, interest-free loans to pay for monthly passes on public transportation and expenses for cyclists.
- Ask for healthy drink alternatives. Companies might increase productivity by encouraging good snacking habits.
- Choose drinks and foods packaged in glass not plastic.

Did you know? Three US medical studies found higher rates of miscarriages among women who were involved in manufacturing semi-conductors than among women in the general population.

- Use cloth-roller towels for bathrooms and kitchens.
- Stock the office kitchen with real paper cups rather than plastic or polystyrene cups, or with real glasses and mugs.
- At lunchtime, if you heat food, microwave it only in ceramic or glass containers, not in plastic ones.
- Ask the person who orders the coffee and tea to switch to organically-grown fair trade products.

CREATE A WASTE-FREE OFFICE

Because some office products contain toxic solvents, stick to those that are odour-free, and use odourless, water-based marker pens and water-based correction fluid. When possible, purchase non-plastic equipment and office supplies. Granted, solid metal paper trays, for example, cost more than their plastic counterparts, but they look good and last forever. Also, make an effort to purchase furniture that was made from sustainably-harvested wood – check that it has the FSC (Forest Stewardship Council) label before you buy. Their website is http://www.fsc-uk.info.

Learn the three "Rs"

Introduce these crucial environmental three Rs to your office:

Reduce

- Distribute circulars rather than individual copies.
- Don't buy over-packaged products; tell retailers that you prefer simple biodegradable packaging.
- Don't over-purchase; encourage co-workers to use up office supplies and order only what they need.

Recycle

- Set up recycling bins for different types of glass, plastics, paper and cans, as well as a box for items to donate to local schools or other businesses.
- Recycle your old computers, or donate them to locate schools. If your community has a literacy programme for immigrants, participants may welcome an older but still functional PC.
- Use fax machines that print on plain paper rather than chemically treated, non-recyclable thermal paper.
- Refill laser, inkjet and fax cartridges or return them for recharging, recycling, or re-manufacturing.
- Have your printing done on recycled paper with soy ink.

Reuse

- Reuse polystyrene 'packing peanuts', but encourage companies to use other, less harmful forms of packaging.
- Reuse boxes and packing materials or find someone who needs them.
- Whenever possible, reuse envelopes, too. Reuse labels are available from your favourite environmental organization.
- Lease equipment or try out someone else's before you buy.
- Look first at used furniture and equipment and ask about re-manufactured copiers, cartridges and other products.
- Purchase products than can be reconditioned or refilled – toner cartridges for laser printers and photocopying machines.
- Use the second side of discarded office paper and make printing on both sides of paper a standard practice.
- Reuse envelopes, files, ring binders and folders, especially for internal use and for occasions when presentation doesn't matter.
- Use string, rather than layers of plastic tape, to tie packages.
- Use rechargeable batteries and green cleaning products.

Did you know? In the US, the world's largest consumer of paper, only five per cent is made from recycled pulp.

CREATE A STRESS-FREE OFFICE

Corporate health programmes are a positive trend in workplaces. These offer healthful cafeteria menus, exercise and sports facilities, and even classes in stress management and yoga. Benefits to a company include an increased sense of teamwork, improved morale and reduced absenteeism. If your company is big, you could suggest – or help start – a programme encouraging people to cycle to work, a sport or running club and a policy to include more organic and vegetarian food in the cafeteria.

Here are ways to improve your office space:
- Green up – literally – the office with real plants. They are good for the air and for the spirit. Increase the humidity in the office by standing plants in trays of pebbles and water.
- Use natural materials such as rattan and wood.
- Check for sources of air pollution, including carpets, photocopiers, art supplies and other equipment.
- Declare your office a no-smoking area.
- Use low-energy light bulbs in ceiling and wall fixtures.
- Make the best use of natural daylight and natural ventilation.
- Ensure good ventilation and that windows can be opened.
- Install ceiling fans – they consume a fraction of the energy of air conditioners.
- To increase natural daylight for employees, position desks by windows. The use of full-spectrum bulbs can also help – spending seven hours a day or more under fluorescent lighting is not good for anyone.

- Position laser printers and photocopying machines at a distance from workstations because they contain toners and solvents that workers should not breathe.

WAYS TO SAVE ON ENERGY

The best step companies can take to help the environment is to cut back everyday commuting and airline travelling. Encourage employees to live near their workplace by introducing incentives to use public transport and to carpool, cutting car-related perks, and supporting telecommuting by providing key employees with equipment to set up a home office.

Here are other energy-saving tips:

- Do not heat or cool unused areas of a building. In holidays and at weekends, set the heating or air conditioning lower or higher, or turn it off, depending on the season. Cut temperatures in work areas by a degree or so; overheating reduces alertness.
- As much as possible, use natural daylight and compact, energy-efficient fluorescent bulbs.
- If you or your company has enough control over your building, draught-proof doors and windows and insulate walls.
- Co-ordinate packages sent by courier, and, if you order by mail, order in bulk. Whenever possible, use fax and e-mail, as well as regular mail, rather than a courier.
- When computers and other equipment are not in use for a period of time, turn them off.

HOW TO SURVIVE THE ONLINE LIFE

Out of 6 million VDU users in the UK, only one in five have exercised their right to an eye test paid for by their employers. Data entry clerks, stockbrokers, bank clerks, secretaries, writers and physicists all spend their days in front of computer screens. Lower-paid workers are even more likely to spend unbroken time working in front of a computer screen and generally have less freedom to take rest breaks. Sometimes the computer directly monitors their output, increasing stress and making it even more likely that they will not take frequent rest breaks. Some campaigners suggest that no employee should spend more than four hours, or half the working day, at a computer screen, and that computer monitors should be properly shielded.

Several companies have policies stating that employees who use computer screens for more than 26 hours a week must take, say, a 15-minute break every three hours. They also provide mobile keyboards, antiglare computer screens and diffuse overhead lighting. But some researchers insist that computers are completely safe and that the symptoms reported by modern office workers around the world, such as dry eyes, headaches and major fatigue, are caused by personal factors or other environmental hazards. Other researchers insist that symptoms such as eye irritation, headaches and muscle aches – commonly called 'repetitive strain injury' – are directly caused by the constant use of computers, combined with poor lighting, poor posture and the physical setup of computer workstations.

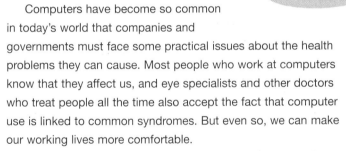

Computers have become so common in today's world that companies and governments must face some practical issues about the health problems they can cause. Most people who work at computers know that they affect us, and eye specialists and other doctors who treat people all the time also accept the fact that computer use is linked to common syndromes. But even so, we can make our working lives more comfortable.

Years ago, typists sometimes suffered sore necks, but they didn't suffer repetitive strain injury, RSI, which is the debilitation that results from performing the same tiny movement for hours on end. One form of RSI, called 'carpal tunnel syndrome' for the carpal nerve in the wrist that is compressed, afflicts millions of people who spend their workdays using a keyboard or mouse – as well as chicken pluckers, tailors, seamstresses, farmers and other people whose work requires them to repeat the same small hand motions over and over again.

The symptoms of carpal tunnel syndrome are hand pain and a tingling in the wrists. Unfortunately, it often strikes with little warning after a long bout of intense keyboard work. Consequently, it's important to take preventative measures before you feel any symptoms. Forward-thinking employers are willing to help workers arrange their workstations to prevent such injuries, as described below.

Eye strain is the problem that computer users report most frequently. Other common problems include upper body tension and lower back pain. It is essential that you adjust your computer screen, as well as your keyboard, to your body. Several manufacturers now make keyboards, mice and mouse pads that create less strain on users, while the new flat plasma screens are easier on users than cathode-ray tube screens.

Unfortunately, we don't all have access to these RSI-reducing improvements right now, so we need to find ways to decrease machine fatigue through good work habits and thoughtful arrangements of our work space.

How to set up your best working position:
- Position yourself 45 to 60 centimetres from your screen with your eyes 15 to 20 centimetres above the centre of the screen.

- Your chair should support the natural curve of your back. Your mother was right: Sit up straight!
- Position your fingers to be level with, or below your wrists. Your wrists should be level with, or below your elbows.
- Have your knees level with your hips so your thighs are parallel to the ground. Rest your feet on a book or low stool if necessary.

Arranging your work station and computer properly:
- Try standing up. A number of famous writers preferred to type standing up, and this stance can relieve back problems. Keep one foot on a very low stool. Standing makes you more mobile in general – you'll move around more frequently.
- Use an old manual typewriter to write personal letters or a diary. Typing on a manual typewriter is great exercise for the hands – and it's fun.
- If you are bothered by overhead fluorescent lights, use a desk lamp with a 100-watt, energy-efficient bulb.
- Eliminate glare by changing the screen position or using an antiglare screen.
- Eliminate static by using an antistatic mat, a bowl of water and plants, or a water spray.

And don't forget your daily routine:
- Vary your work. Remember to switch from task to task at regular intervals. Take a break at least every half-hour. Put down the pencil, pen or mouse, stand up, and move around.

- Blink every three to five seconds – computer users tend to forget to blink, and thus their eyes become dry – and use eye drops if necessary.
- Look away from your computer screen frequently – at least once every five minutes.
- Move your eyes. Look out the window or even at a picture on a distant wall. You might want to put a mirror near your screen to create additional visual distance.
- Use keyboard commands instead of mouse clicks – for example, Ctrl+C=copy, Ctrl+A=select all.
- Clean your computer screen and your contact lenses or eyeglasses frequently.

Get moving!

Exercise can help you prevent many health problems. Yoga, in particular, often alleviates – or even prevents – the symptoms of carpal tunnel syndrome. More specifically, Iyengar yoga contains a variety of excellent hand and arm postures, but any yoga practice is likely to help.

Exercises you can perform at your desk:
- To alleviate general aches and pains throughout the day, roll your head in full circles whenever you get a chance. Stand up and stretch both hands above your head, pressing your feet hard into the floor.
- For relief of neck strain, massage your temples and the base of your head above the neck.

- To relieve a headache, close your eyes and rub your earlobes, squeeze the bridge between your eyes, or press on the fleshy web between your thumbs and first fingers.
- Close your eyes, relax your face and brow, and move your eyes from side to side as you breathe deeply.
- Exercise your eyes by shifting them in corner-to-corner patterns on your computer screen.
- To release tension and wake up your eyes, inhale deeply and squeeze your eyes shut as tightly as you can while tightening your jaw, neck and face. Exhale quickly while stretching your mouth and eyes wide open.
- To release tension in your hands and fingers, drop them to your sides and shake them for 15 seconds. Then extend your fingers as far as you can a few times and rotate your wrists from front to back.

Our jobs and careers can be a great source of satisfaction, and workplace communities are vital social networks. Environmentally responsible companies – whether they're directly involved in environmental products and services or just supportive of green policies – are essential players in making our world a better place.

CHAPTER SIX

GREENER TRANSPORT, SAFER PLANET

Motor vehicles are among the most damaging of human inventions. But they've become a major part of our lives, and we're no more likely to give them up than we are to give up computers and electric lighting. So let's lift the bonnet, look the problem squarely in the motor, and see what we can do about it. This chapter examines what's wrong with our dependence on the car. It also explores ways to reduce our driving, improve our health with greener mixed transportation systems, neighbourhoods that are not totally centred on cars, and, finally, holiday travel.

Greener transport isn't just a goal for us to aim towards because we want to be virtuous. It will have real benefits almost immediately. It will make society healthier, communities stronger and more pleasant to live in, and medical bills lower.

Burning issues

We've become accustomed to the convenience – and the mindlessness – of driving. But somewhere along the way, we've also become oblivious to the many inconveniences and negative effects. It seems that the 'tragedy of the commons' effect is operating here. Ecologists came up with this term to describe a situation in which a natural resource – a lake, a forest, or even a village green – is owned in common by a group of people, but because there are not sufficiently strong social structures to make sure that the resource is maintained, it is allowed to run down. No one owns it, so no one takes responsibility for it!

Much of the time we don't really like other people in cars – the rudeness of other drivers, the noise, the exhaust fumes, the traffic jams – or the effect that cars have on our communities. Lots of people even try to live in places where they can get away from cars. But cars are the way of the world, and car ownership is considered to be the most important sign of affluence. Because vehicular transport contributes one-third of the carbon dioxide that causes global warming, many experts believe that our main environmental imperative is to rethink the way we get around.

Did you know? Burning just 5 litres of petrol or diesel fuel releases between four and nine kilograms of carbon dioxide into the atmosphere, the actual amount varying according to the fuel efficiency of the engine.

Here's how to make your car a little easier on the environment:

- Share and share alike – organize car pools to travel to work.

- The next time you buy a car, choose one that emits fewer pollutants and gets more mileage out of a litre of fuel.

- Plump up at the pump. Eighty per cent of tyres are under inflated. By keeping your car's tyres properly inflated, you can save money at the pump, make your tyres last longer and help the environment. If all drivers would keep their tyres properly inflated, we could save more than 35 million litres of fuel every day.

- Buy low-rolling-resistance tyres. The Michelin Energy is a green tyre that uses special tread designs, lower-profile side walls, higher pressure and lighter materials to improve fuel efficiency and reduce resistance without a loss of handling.

- Less cool, less fuel. When we drive slowly, using the car's air conditioning can reduce fuel efficiency by more than 20 per cent. When we drive faster than 64 kilometres an hour, however, using the air conditioning is actually more fuel efficient than rolling down the windows, because it does not create as much aerodynamic drag as open windows do.

- Combine trips if possible, and share a single car between your family. Getting by with just one car can lower your fuel, insurance and maintenance costs by hundreds of pounds.

- What's up, Doc? Your petrol bill is what's up if you make bunny-hop starts. They burn twice the fuel that is used when you set off in smooth, gradual style.

In developed countries, the proportion of the transport budget allocated to the car is huge. People are encouraged to own cars because of the roads that are built, tax concessions for road building, and a general neglect of funding and programmes to support public transport. New shopping centres and malls continue to draw shoppers away from town and local centres and into out-of-town locations that are geared to car owners – in spite of the fact that in the United States, where this trend started, 20 per cent of malls that were in operation in 1990 have closed.

Today, more Americans are looking for places where they can shop and dine – and often live, too – without having to use their cars. Consequently, traditional high streets are being revitalized, and neo-traditional developers are building shopping areas and communities on the old model. We can learn from that example.

Sick of driving

Car engines are a major cause of air pollution because they release carbon monoxide, nitrogen oxide, benzene, hydrocarbons, sulphur dioxide and lead. These pollutants can cause headaches, burning eyes and throat, as well as various long-term respiratory and blood problems, including bronchitis, asthma, lung cancer and leukaemia. The annual global health cost of air pollution is estimated at £13 billion.

Driving is equally harmful to our health because it requires no physical effort. We can't burn a lot of calories gripping a steering wheel and pressing the pedals. Traffic noise and congestion also create a great deal of stress today. Many people will go to almost

any length to avoid getting caught in traffic jams, and reports of so-called road rage are increasing as drivers' frustrations boil over into violence toward pedestrians, cyclists and other drivers.

Communities and our cars

Cars are inherently dangerous. They are large, heavy objects that travel at high speeds in close proximity to other cars, cyclists and pedestrians. They take up large amounts of space – as much as one-third of the surface area of a town – and dominate our physical space. They have dramatically restricted our children's freedom to roam, making them less fit and less independent.

Researchers increasingly find that when schools, shops, workplaces and homes are separated by long distances, people meet only in impersonal and well-defined indoor spaces, not in the casual spaces that make us feel connected to each other. This finding reveals an important element of our modern sense of alienation and a reason for the increased interest in neo-traditional housing developments – those designed for walking and so promoting human interaction. In these developments, parking space is often shared and positioned out of sight.

Out of touch, out of mind

The less contact we have with a place, the less we care about it. When you read that the countryside is threatened by development, it doesn't mean much to you from the front seat of a car that's speeding down the highway. But highways gouge wide scars through large stretches of the pristine countryside, obliterating everything in their path. En route to our destinations, we city dwellers can be extraordinarily insensitive to the ill effects of our demands on even nearby stretches of countryside and can let our convenience override the destructive impact that building more roads has on small towns and rural areas.

For the sake of more roads, many valuable and beautiful wildlife sites have been destroyed and more are facing the bulldozer each year. Predictably, roads and road-building have a disastrous effect on animals. A strip of hard, wide tarmac populated by continually moving, high-speed metal missiles, poses a barrier to animals that is at the very least dangerous and at most insurmountable. Road noises will generally drive animals away, and toxic exhaust fumes pollute their food and water. Scientists consider roads to be a major threat to biodiversity, and roads themselves are changing the character of communities and natural landscapes all over the world.

Imagine... Your street with only a quarter of the cars that are now parked there. All of that free space could be filled up with trees, seating areas, flowering shrubs and safe play areas for the children.

TIPS FOR IMPROVING TRAFFIC

Traffic patterns and parking regimes are two huge factors in creating livable communities. Here are six tips to remember as you try to improve the traffic problems in your town or locality:

- *Work to improve public transport and increase public awareness of the benefits of using it.*

- Make residential streets unattractive to rat-racing motorists by adjusting traffic priorities at junctions, using one-way systems and roundabouts and closing narrow roads to traffic.

- *Reduce speeds with chicanes and 'sleeping policemen'.*

- Encourage home-owners and the local council to plant shrubs and trees, which improve the air, muffle traffic noise, and make streets much more pleasant for residents.

- *Exempt cyclists from some one-way restrictions and road closures. Provide cycle paths and turning lanes.*

- Demand frequent pedestrian crossings and wider and higher pavements to increase pedestrian safety.

- *Ask for well-lit pavements and pedestrianized shopping areas. They encourage people to leave cars at home.*

TIPS FOR DRIVING LESS AND SAVING MORE

* *Keep your tyres fully inflated and your engine tuned.*

* Ensure that your car's battery and oil are recycled.

* *Join the Environmental Transport Association at www.eta.co.uk.*

* Use radial tyres and asbestos-free brake pads.

* *Remove roof racks not in use. Empty the boot regularly.*

* Plan ways to minimize car trips. When you shop, buy in bulk – this is convenient and saves money, too – and do several errands on one trip.

* *Don't drive alone. Car-pool to work if you cannot use public transport; share driving the children to school; and do your weekly shopping with a friend.*

* Tune up your bike and start wearing comfortable walking shoes.

* *Try not to use the car for short trips – catalytic converters don't work until they're warmed up.*

* Purchase a second-hand car. Major mechanical problems generally show up during the first 30,000 kilometres, so second-hand cars can be better buys than new ones.

* *Avoid diesel engines – they produce much more particulate pollution than petrol engines.*

* Slow down to save fuel. Fuel consumption is highest in stop-and-start town traffic and – depending on the car – at more than 85 kilometres per hour on the highway.

* *Burn greener fuels – natural gas, methane, or ethanol.*

* Buy a hybrid car that uses solar energy as well as petrol.

* *Always drive with patience and control – you'll even save money! Aggressive driving consumes 20 per cent more fuel. Avoid slamming on the brakes, never rev the engine, and use the gears to slow down.*

* Try to organize your life so that you can live in an area where you won't have to drive as much.

* *Live as close as possible to your company's office, work at home or talk to your company about being allowed to telecommute one day a week.*

The road to remedy

Our transport problems have no single solution. However, the first consideration of town planners and government officials should be to provide accessibility, not necessarily mobility. Our goals are to reach certain places and do certain things. One good rule of thumb is to use a car only when we are making a journey of more than three kilometres, and we should try to live within that radius from the goods and services we need. The evolution of global communications makes it possible for us to reduce car trips to some extent, and if we adopt a mixed transportation strategy for ourselves and our households – using public transport, walking and cycling, and driving only if we must – we can keep literally tons of carbon dioxide out of the atmosphere.

We drive everywhere, so it's no wonder obesity is becoming an epidemic. But being active and outdoors – in rain, wind and even snow – is good for us. Our bodies are designed to cope gracefully with most weather and it's invigorating to experience the seasons. Seasonal temperature changes are good for your immune system, so don't think that you have to be a shut-in in deep winter, especially when wonderful high-tech athletic gear is available and works well for all-season runners or cyclist commuters.

The 'u' and 'i' in 'public'

Both city people and country folk need more reliable, cleaner, faster, safer and more comprehensive bus and train services. When country bus routes and branch lines were closed, many people in rural areas become either isolated or even more dependent on their cars.

But people won't turn *en masse* to public transport until the government puts money into improving services. Government expenditure on public transport is called a subsidy, while the much greater expenditure on more and bigger roads is not. We should support public transport, both by voting for it and by using it whenever we can. We should also join public pressure groups as well as local associations that campaign about traffic issues. And remember that the best way to speed up driving is to improve public transport.

Just consider a few of the personal benefits of not driving. On public transport you have time to doze, meditate, read, or eavesdrop, with no worries about having a drink or having to find a place to park. However, travelling by public transport does require a different level of advance planning from travelling by car:

- Note the times of the last trains, buses and underground trains. Keeping the phone number of a reliable taxi service handy isn't a bad idea, either.
- Ask your employer about working on flexi-time – this arrangement lets you travel at off-peak traffic times.
- Assemble a complete collection of timetables and maps for your area, or bookmark websites that will help you plan local and longer trips.
- Prominently note the information numbers or web addresses of services that you use often.
- Inquire about lower fares for off-peak travel or cards that bring you reductions.

Walking – the first step

Walking is the single best activity to support our mental and physical health. For example, women with breast cancer who walk moderately each day have a 25 per cent greater chance of recovery. This fact might seem remarkable, but just consider this: during all the millennia before the internal combustion engine, our foremothers walked everywhere, all the time. Our bodies are built to walk; our bodies need to walk, and to walk a lot.

Pragmatically, walking is a way to get from point A to point B, but it is so much more. It's a vital and guiltless pleasure, the best way to refresh your spirit, and an ideal background activity for concentrated – or idle – thought. With your feet on the ground instead of the clutch, brake and accelerator, you notice the first tinge of autumn colour in the trees and enjoy the scent of newly-turned earth or a hedge of may coming into blossom.

Walking lets you enjoy the health benefits of breathing fresh air and basking in natural light. You don't need any special gear, but carrying a small backpack can be more comfortable than carrying a handbag. Put your glamour-girl shoes in the backpack and change into them when you reach your destination.

Life cycles

Cycling is an enjoyable, efficient and convenient way to get around. It's much faster than walking, faster even than driving in many cities, and eminently suited for short trips.

In spite of the inevitable frustrations that cyclists suffer by sharing the road with motorists, cycling is an important element

of a mixed transport system. Worldwide, bicycles outnumber cars two to one. The bicycle is a vehicle for people of all ages and is sometimes called the 'vehicle for a small planet'. But bicycle use varies greatly. For example, Britain lags well behind several of our European neighbours in cycle friendliness, as the packed bicycle racks outside train stations in European cities show. In Denmark, people make ten times more journeys by bicycle than people do in Britain – 20 per cent compared to two per cent in the UK. In Amsterdam, where the climate is no better than in much of Britain, one-third of all trips are made by bicycle, and in the city of Gröningen, more than half the population travels by bicycle.

The more people cycle, the safer all cyclists will be. We need dedicated bicycle paths, more marked bicycle routes and counterflow lanes on one-way streets. In the UK almost 13,000 kilometres of cycle paths, and more than 11,000 kilometres of signed cycle routes, are operated by Sustrans, a charity devoted to decreasing car use and promoting walking, cycling and public transport. For long distance and local cycle route maps and news of national and local cycling events try www.sustrans.org.uk.

If you would like to start bicycling but feel nervous about it, stick to local trips and quiet streets until you develop the confidence and skills to cope with motor traffic. Bicyclists can be far more flexible than drivers, using quieter residential streets to make travelling faster, safer and less stressful. Friends of the Earth can put you in touch with local cycling groups, many of which distribute maps of good routes and cycle paths. You can take bicycles on most trains and sometimes even on buses.

Tykes and bikes

One-quarter of the trips made by families with children under 15 years old are 'escort trips' – taking children to and from school, friends' houses, sports events and practices or music lessons. Is it any wonder that our children don't get enough exercise?

Cycling is a great way for older children and teenagers to get around, and is also a great way to develop physical co-ordination and self-confidence. Remember the challenge of mastering the high art of riding on two wheels, and the thrill of achievement?

Help your children plan cycle routes and take trips together. You can even carry a young child on your own bike, using a child seat and child-size helmet. Some parents use a tricycle or a trailer seat, which is also very practical for shopping.

How to be well-wheeled

- Attach a big basket to your bike to make carrying things easier.
- Wear a pair of old running shoes – bicycle pedals and toe clips can chew up office shoes – and carry an extra shirt if the weather is warm.
- Buy a second-hand bike. You don't need a rugged mountain bike or a sleek, ultra-light racer for commuting or shopping.
- Be sure that the bicycle seat is at the right height and is comfortable for you. Women may want to buy a seat that is made to accommodate a wider pelvis.
- Wear plenty of reflective gear, a helmet and clips or bands on your trouser legs.
- A bicycle bell is a simple and very effective safety device.

TRANSPORT OF THE FUTURE

At this late date, nothing is going to eliminate the negative social effects of cars or make them take up less space. And we clearly are not going to return to a carless society. Therefore, we need brilliant innovations. We need cars with super efficiency, new materials, direct-hydrogen fuel cells, integrated whole-system engineering – cars that go far beyond today's green cars.

The 2004 Toyota Prius gas-electric hybrid releases half of the pollution and burns half the petrol of a mid-size conventional car. And researchers are developing designs and technology to make a hybrid electric car that is safer to drive, roomier, and more competitively priced – a car, according to energy guru Amory Lovins, that "emits nothing but hot drinking water, doesn't rust or dent or fatigue, performs like a sports car, and gets 110 mpg." Such cars will reduce the carbon dioxide problem that current models create by about two-thirds. Innovative people are making serious efforts to get them on the market, and you can help by buying them when they become available or by investing in their development now.

HOLIDAYS AND TRAVEL

The greenest holidays are those you take at home – travelling to other parts of the country or the world uses lots of energy and resources. But we can learn a lot by travelling, and people who travel sensitively are often the most alert and active global citizens. Environmental problems are global, and travel, at its best, can help us learn to work together to solve problems together. It's crucial to be aware that tourism is now the world's largest industry, with annual revenues of more than £220 billion. On a global basis, it employs one in 15 workers. But aeroplanes, touring coaches, cross-country trains, recreational vehicles and cars pulling caravans contribute tons of carbon dioxide to the atmosphere each year, along with a considerable percentage of the total of climate-altering gases. In addition, 60 per cent of ozone-destroying gases come from aircraft – and the percentage is rising.

The tourism and travel sectors are among the fastest-growing sectors of the economy in many of the world's countries, both rich and poor. Therefore, we must find ways to minimize the damage that tourism and travel inflict and maximize the social and economic benefits that they can bring.

Did you know? The number of tourists worldwide is expected to triple in the next ten years, to an annual billion people. Revenues will quadruple, but what of the pollution?

The irony of eco-tourism

Many of us go on holiday in search of pure escape – peace and quiet, with majestic mountains or surging seas by day and starlit skies by night. But if millions of us travel the globe in search of secluded retreats, eventually no place will be remote, and we won't be able to find a quiet spot. We take our world with us when we travel, and all too often we destroy the thing we love. In fact, a recent World Wildlife Fund (WWF) study found that some of the most profitable tourist destinations – Florida, Spain and the Alps – are likely to be dramatically affected and perhaps even ruined by global warming.

All over the world, investors and visitors are destroying the vital differences that mark out life in various regions. Local customs and religious rituals are transformed into spectacles, and traditional craft objects become cheap souvenirs. Ecosystem balances are being disturbed by mountain chalets, ski areas and beach developments. Even wilderness areas that people can reach only on foot are being littered and eroded. Community and family ties are being broken as residents of tourist areas become increasingly dependent on travellers for their livelihood.

Tourism does *not* bring prosperity to poor regions. The tourist industry generates only unskilled jobs for residents – skilled jobs tend to go to foreign employees – and 80 per cent of the money that travellers spend comes back to the home countries in the form of foreign staff salaries, agency commissions, foreign-owned hotel profits, payment for imported food and other items, insurance and interest on loans. A few wealthy locals and officials reap most of the remaining 20 per cent.

Cruises represent travel at its western worst, polluting and consuming without concern for the environment – cruise ships dump sewage and garbage into the sea – and treating local cultures as something to be consumed rather than truly experienced. Adjusting to a different kind of travel can be difficult, but well worth it.

Travelling is never more exhilarating and enriching than when you really engage with a place and people. Many people travel to learn – about other cultures, history, and the natural world. At its best, travel helps us to understand our global village and to appreciate the material wealth that so many of us enjoy.

Getting ready to go

- Learn at least a few useful phrases in the languages of the countries that you will be visiting.
- Remember that each bit of weight that you take with you – on a plane or in a car – requires extra fuel to move it along. So don't try to get around airline weight restrictions.
- Pack and wear gently worn clothes that you can give to someone before setting off for home.

A MOVABLE FEAST

Most of the food you can find when you're travelling is heavily processed, unhealthy and over-packaged. You can enjoy better food and save money by keeping basic picnic supplies in your glove compartment or overnight bag. Pack salt and pepper, a corkscrew, a few pieces of cutlery, and small hand towels to use for napkins and cleaning up. During the day you can stop at a shop or market to buy additional provisions.

Eat local specialities and keep a lookout for independent, locally-owned small restaurants off the main tourist routes. See www.slowfood.com for details.

HOTELS

Hotels consume huge amounts of energy, water and materials, but some realize that they can improve public relations – and save millions – by following 'green hotel' policies. These include installing energy-efficient light bulbs and heating systems, water-saving devices and body-care product dispensers instead of miniature bottles. They display cards offering guests the choice of not having their bed and towels changed daily, and remind guests to turn off the air conditioning when they leave their room.

WORKING HOLIDAYS

One of the current mantras of educators and even politicians today is 'lifelong learning'. All of us need to continue to learn and one way to do that is by taking working or service holidays. Service holidays can give you a real change from routine as well as a sense of how other people live. For example, you could visit an organic farm and help with planting or harvesting. Check www.wwoof.org for farms welcoming volunteer workers worldwide.

We can also use holidays to explore new options. Can we make a living and find a sense of purpose at the same time? Recent surveys show that college graduates are thinking about careers in terms of meaning and satisfaction – making a difference. You might want to go into politics and work for more sustainable economic policies, into environmental law or ethical investment, and use a holiday to gain some experience of your chosen field or awareness of career alternatives.

MEET THE PEOPLE

Around the world, sincerity and genuine interest are quickly recognized and appreciated. The key to having a true local experience is to approach it in manner that fits your lifestyle. Some travellers prefer to stay in familiar tourist hotels, eat in tourist restaurants, and visit only the major attractions. Others go to the other extreme and try to go native, by living as much as possible as if they were locals.

For most people, a more middle-ground approach that combines genuine involvement in a local culture with Western amenities brings the richest experience. Prepare for your trip by reading about the places you will visit and, if you can, by talking to someone who is familiar with them. Standard guidebooks aren't much help when it comes to culture, and you may get more information from current non-fiction books or even novels set in the places you will visit. Check the Web, too, for information that goes beyond the usual lists of hotels.

TIPS FOR THE GREEN TRAVELLER

Learn about where you are going. Buy or borrow a good guidebook or a novel set in the area.

Take a train rather than a car or an aeroplane. Could you travel some of the way by foot or by bicycle?

- *Choose a destination in which you and your family are genuinely interested instead of 'doing' the sights.*

- Find a connection with a local person – through a colleague, an old friend, or the Internet. Many people are thrilled to be able to show their home to a visitor.

- *Try a home exchange (see www.homeexchange.com) – you'll save money and and have a richer experience.*

- If you do use a tour company, choose one that is sensitive to ecological and ethical issues.

- *Rather than look for a safe, home-away-from-home experience in a Western-style hotel, be more daring and experience something different.*

- Don't overuse towels and hot water just because you're not paying extra for them.

- *Leave those little bottles of shampoo at the hotel. They use too much packaging and you don't need them.*

- When you leave your room, turn off the lights and turn down the heating and air conditioning.

Making memories

- As you tour, choose one or two wonderful souvenirs, buying directly from a craftsperson if you can, rather than from a hotel shop.
- When you're packing to return home, remember that golden rule of wilderness travel: "Take nothing but photographs and leave nothing but footprints."
- Don't go overboard with photographs, either. Conventional film processing is a polluting business, although eco-labs recycle canisters and film cassettes, recover silver from film, and flush no chemicals down the drain.

THE AIR UP THERE

Airlines face intense competition to keep profits high, and many flights these days are packed full, with seats jammed ever closer together. Worse still, airline passengers are suffering oxygen deprivation as airlines cut costs by using ventilation systems that draw in less than 50 per cent fresh air, recycling the rest of it throughout a flight. Recycling air makes an aeroplane more energy-efficient, but it also jeopardizes passenger health. The Aviation Health Institute claims that cabin air quality has declined so much that passengers suffer, in effect, from carbon dioxide poisoning. No one knows how stale the air is on commercial flights. But clearly, we need to encourage governments to monitor air quality standards in the hope that we won't have to resort to travelling with oxygen inhalers.

WING TIPS

- *When you arrive at your destination, spend as much time as possible in daylight. Doing this will enable your internal clock to reset itself, and you'll recover from jet lag more quickly. Sex helps, too!*

- Wear a good quality, effective, sunblock.

- *Avoid dehydration. Drink lots of water, avoid alcohol and caffeinated drinks, and bring along a spray bottle of water to keep your skin moist.*

- Breathe deeply to increase your oxygen intake.

- *Take shorter flights. And if you have a chance to get off the plane, do so – jog or walk around the airport, or go outside for some fresh air.*

- Order a vegetarian meal when you make your reservation.

- *Refuse all over-packaged items and write to the airline to suggest better environmental policies.*

AFTERWORD

From a biologist's point of view, it's not our planet that needs to be saved. The Earth will take care of itself. You and I, though, we need saving if we want to have a place to call home.

Here are four easy principles to keep in mind:

- Small is beautiful. Small cars, small houses, and small pets have less environmental impact, and are easier to maintain.
- Don't sweat the small stuff. *Do* take care over major buys like a car or refrigerator. Don't agonize over a plastic bag!
- One step at a time. Turn the thermostat down just one degree, not ten. Then another degree, and another.
- Be a leader if you can. Be among the first to adopt new technologies such as hybrid cars. This is what creates the buzz needed to take important innovations into the mainstream.

It is important, too, to share your values with others, in a gentle way. Showing is almost always better than telling. Offer to bring a picnic basket and pull out real, non-disposable plates (picked up at a jumble sale) and cloth napkins. Turn up at a party on a bike.

I've shared many of my ideas in this book, but there are more coming all the time – what works for me may not work for you, and something you come up with may be just the advice I need. We need to build a global community, to make sure good ideas are passed around. We need to support one another and join together to fight developments that will harm the world we share.

INDEX